Douglas Carnegie, Matthew Moncrieff Pattison Muir

**Practical Chemistry**

A Course of Laboratory Work

Douglas Carnegie, Matthew Moncrieff Pattison Muir

**Practical Chemistry**
*A Course of Laboratory Work*

ISBN/EAN: 9783337279011

Printed in Europe, USA, Canada, Australia, Japan

Cover: Foto ©berggeist007 / pixelio.de

More available books at **www.hansebooks.com**

# PRACTICAL CHEMISTRY.

London: C. J. CLAY AND SONS,
CAMBRIDGE UNIVERSITY PRESS WAREHOUSE,
AVE MARIA LANE.

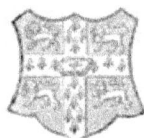

Cambridge: DEIGHTON, BELL, AND CO.
Leipzig: F. A. BROCKHAUS.

# PRACTICAL CHEMISTRY

## A COURSE OF LABORATORY WORK

BY

## M. M. PATTISON MUIR, M.A.

FELLOW AND PRÆLECTOR IN CHEMISTRY OF GONVILLE AND
CAIUS COLLEGE,

AND

## DOUGLAS CARNEGIE, B.A.

DEMONSTRATOR IN CHEMISTRY AND FORMERLY SCHOLAR OF
GONVILLE AND CAIUS COLLEGE,

CAMBRIDGE.

A COMPANION-VOLUME TO PATTISON MUIR AND SLATER'S
ELEMENTARY CHEMISTRY.

CAMBRIDGE:
AT THE UNIVERSITY PRESS
1887

𝕮𝖆𝖒𝖇𝖗𝖎𝖉𝖌𝖊:

PRINTED BY C. J. CLAY, M.A. & SONS,

AT THE UNIVERSITY PRESS.

# PREFACE.

THIS book forms one part of a course of elementary chemistry; the other part is contained in the companion-volume entitled *Elementary Chemistry*. The two books are intended to be used together, the one being complementary to the other.

The third part of this book deals with subjects which are only touched on in the companion-volume; this part should be used in conjunction with portions of the *Principles of Chemistry* by one of the authors of the present book; references to original papers to be consulted are also given in this part.

The aim of the authors has been to arrange a progressive course of practical chemistry, in which as the experiments become more difficult the reasoning becomes more close and accurate. The arrangement of the course and the selection of the experiments are the outcome of experience gained in teaching chemistry for many years. Most of the experiments described can be performed with the apparatus to be found in every moderately well-furnished laboratory.

To encumber a new book on practical chemistry with the details of qualitative and quantitative analysis—subjects treated so fully in numberless manuals—appeared to the authors to be unwise. Many of the experiments in the earlier chapters of this book indirectly teach the principles of qualitative analysis, but the authors' plan has obliged them to assume that at certain stages of his progress the student has become acquainted with the ordinary processes of qualitative and quantitative analysis.

Appendices are added, containing (1) the outlines of experiments bearing on the work done in Part I., (2) tables which may help the student in performing the analytical parts of the easier experiments, (3) numerical data frequently used in the laboratory.

M. M. PATTISON MUIR.
DOUGLAS CARNEGIE.

CAMBRIDGE, *October*, 1887.

# TABLE OF CONTENTS.

## PART I.

## PART II.

# PART III.

# APPENDICES.

# COURSE OF PRACTICAL CHEMISTRY.

## PART I.

## CHAPTER I.

### CHEMICAL AND PHYSICAL CHANGE.

**Exp. 1.** Hold a piece of *platinum wire* in the flame of a Bunsen-lamp for a few minutes. Then hold a piece of *magnesium ribbon* in the same flame.

The platinum becomes red-hot, glows, and emits light; when removed from the flame, it presents the same appearance as before it was heated. The magnesium also glows; but in addition to this, it is burnt: a white powder is produced, unlike the magnesium; this white powder is called *magnesia*.

The change produced in the platinum was a physical change. The change of magnesium to magnesia was a chemical change.

**Exp. 2.** Magnetise a *knife-blade*, by drawing the poles of a horseshoe magnet over it several times in the same direction. Bring the magnetised blade close to a quantity of iron-filings; the iron is attracted to the blade.

The steel which forms the blade has acquired a new property, but it still exhibits all those properties which characterise steel.

Place some *iron-filings* in a porcelain basin, and add some dilute sulphuric acid; when effervescence has ceased, add a little more acid, then evaporate until the liquid becomes slightly thick, but take care that the whole of the iron has not disappeared ; filter while hot, and allow the greenish coloured liquid which runs through the filter—*the filtrate*—to cool.

Green crystals of *sulphate of iron* are formed as the liquid cools. Compare these crystals with the iron-filings with which you began the experiment, as regards colour, appearance, hardness, and solubility or insolubility in water ; the green crystals are evidently quite a different kind of matter from the iron.

The change produced in the knife-blade was a physical change. The change of the iron to sulphate of iron, by causing it to interact with sulphuric acid, was a chemical change.

*By the process of filtration, a liquid is separated from a solid. The liquid which runs through the filter is called a filtrate.*

**Exp. 3.** Heat a little *iodine* in a large dry test tube, heating only that part of the tube where the iodine is.

The iodine slowly changes to a dark violet vapour, but as this comes into contact with the colder parts of the tube, a solid body is formed ; this solid is easily seen to be the same as the iodine before it was heated.

*The change of a solid to a gas, followed by the re-formation of the solid on cooling the gas, is called sublimation.*

Heat a little dry powdered *lead nitrate* in a dry test tube.

A reddish-brown gas, called nitrogen tetroxide, is formed, and a yellowish-white solid, called *lead oxide*, remains.

The action of heat on the iodine has been to produce a physical change. The change of lead nitrate to lead oxide and nitrogen tetroxide, produced by the action of heat, has been a chemical change.

**Exp. 4.** Dissolve a little *common salt* in water,* in a porcelain basin ; place the basin on the top of a beaker, of a

Fig. 1.

size such that part of the basin is within the beaker ; put a piece of paper between the beaker and basin ; put some hot water in the beaker and place it over a lamp. (Fig. 1.) As the water boils the steam surrounds the basin and heats its contents nearly to 100°. The water in the basin is thus evaporated, without danger of the solid matter, which is formed as the water passes away, being lost by spirting. When the contents of the basin are dry, collect some of the

* Always use *distilled water*.

white solid and compare its properties with those of the common salt you used ; note the colour, appearance, taste, and solubility in water, of each.

*In all cases where evaporation of a liquid is to be continued until the liquid is wholly removed, the final stages of the process should be conducted on a water-bath* (as described above), *unless special directions are given to the contrary.*

The salt has been changed by the action of water; it disappeared in the water ; but by removing the water the salt has been again obtained. The change has been physical, not chemical.

Place a little water in a basin, and into it throw one or two small pieces of a white, soft, lustrous, solid called *sodium.*

*Never touch sodium with wet fingers.*

The sodium slowly disappears in the water with a hissing noise ; when the sodium has all gone, place the basin on a water-bath and evaporate until the water is wholly removed.

*This is called evaporating to dryness.*

You obtain a white, hard, lustreless, solid, very unlike the sodium thrown into the water. This solid is called *caustic soda* or *sodium hydroxide.* The change of sodium to caustic soda is a chemical change.

**Exp. 5.** Mix a little blue coloured solution of *litmus* with some colourless water. The result is a liquid coloured lighter blue than the litmus ; the properties of this liquid are those of the litmus added to those of the water.

Mix the two colourless liquids, solution in water of *potassium iodide* and solution in water of *mercury chloride.*

A reddish-yellow solid is at once formed, called *mercury iodide.* The appearance and colour of this shew that its properties are different from those of either of the bodies by the interaction of which it has been produced ; it is evidently a different kind of matter from either of these.

*A solid substance formed by the interaction of one liquid with another, or of a gas with a liquid, is generally called a precipitate. The mercury iodide is the precipitate in the foregoing experiment. We shall use the contraction pp. for precipitate.*

The litmus and the water were physically changed : no new kind of matter was formed. The mercury chloride and the potassium iodide solutions were chemically changed : a new kind of matter, mercury iodide, was formed.

**Exp. 6.** Dissolve a few lumps of *sugar* in warm water in a beaker : then evaporate the liquid to dryness on a water-bath. Notice the bubbles of air which rise to the surface of the water during solution ; this air was imprisoned in the pores of the sugar, and as the solid sugar dissolved in the water it was set free. After the water has been removed by evaporation you obtain a white, sweet, substance, the properties of which indicate that it is sugar.

Place a few pieces of *marble* in a bottle with a cork and tubes arranged as shewn in Fig. 2. Pour a little water on

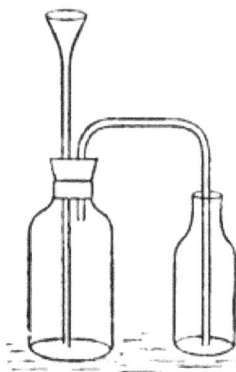

Fig. 2.

to the marble, then pour a little hydro-chloric acid down the funnel tube into the bottle.

The marble gradually disappears in the water and acid, and a gas is produced. Allow this gas to pass into a dry bottle full of air as shewn in the figure. After 5 minutes or so, bring a lighted taper a little way into the bottle ; the light goes out.

You thus prove that when marble interacts with hydrochloric acid and water, a new kind of matter, very unlike either the marble or the acid, is produced. This new kind of matter is a colourless, odourless, gas : you recognise its presence in a vessel by making use of the property it possesses of ex-tinguishing a lighted taper.

This gas is called *carbon dioxide* or *carbonic anhydride* ; it is heavier than air, and can therefore be collected, as you have collected it, by allowing it to pass to the bottom of a vessel full of air.

*The method of collecting gases heavier than air which you have just used is called collection by downward displace-ment. It is often used, especially for collecting heavy gases which dissolve in water.*

The change of sugar into solution of sugar was a p h y s i c a l c h a n g e ; no new kind of matter was produced. The change of marble into carbon dioxide was a c h e m i c a l c h a n g e ; the properties of the matter produced differed in a very marked way from those of the marble.

These experiments illustrate the prominent differences between physical and chemical change. When a specified substance is physically changed, it temporarily acquires a new property or new properties; but the substance is present, and can be recognised by its ordinary properties as being present, during and after the physical change. When a substance is chemically changed, the original substance disappears and at least one new substance is produced in its place. The differences between the properties of the original substance and the new substance are so marked that we do not hesitate to call each a different kind of matter from the other.

The experiments in this chapter have illustrated the meanings of the terms;—*filtration, sublimation, solution, evaporation, precipitation.* They have also shewn how to use a *water-bath* for evaporation, and how to *collect a gas* heavier than air *by downward displacement.*

*Reference to* "ELEMENTARY CHEMISTRY." Chapter I.

# CHAPTER II.

**Exp. 1.** Place a piece of *magnesium* ribbon in a porcelain crucible (about 12 inches of ordinary magnesium ribbon loosely wrapped together), put on the lid, and counterpoise the whole on a fairly good balance against shot or small pieces of metal. Set the crucible on a triangle standing on one of the rings of an iron stand (Fig. 3). Heat the crucible gently, then strongly, and then over a good blow-pipe; raise the lid from time to time, for a few seconds, to admit air.

The magnesium is slowly burnt to *magnesia* (*comp. Chap. I. Exp.* 1). Do not remove the lid at any time for more than a second or so, else some of the magnesia will be volatilised and lost. When the burning is finished allow the crucible to cool, then place it on the balance and the counterpoise on the other pan.

Fig. 3.

The crucible and the magnesia together weigh more than the crucible and the magnesium.

The change of magnesium to magnesia is a chemical change: the product of this change, magnesia, weighs more than the magnesium.

**Exp. 2.** Place a little *very finely divided iron* in a crucible, and counterpoise the whole. Heat the crucible over a lamp until the iron glows throughout. Allow to cool, and counterpoise again. Compare the appearance and action towards a magnet of the substance in the crucible with the appearance and action towards a magnet of the original iron.

There has been an increase in weight : and a new kind of matter, *iron oxide*, has been produced.

Magnesium and iron have been chemically changed, each into a kind of matter different from itself: the new matter produced in each case weighs more than the original matter. Therefore the change has consisted in the addition to, or combination with, the iron, and the magnesium, of some other kind, or other kinds, of matter.

**Exp. 3.**   Place a little *finely divided copper* in a piece of hard glass tubing about 6 ins. long. Counterpoise the tube and its contents. Connect the tube, by means of a cork and glass tube, with a U tube containing calcium chloride, which is a substance that quickly absorbs moisture (Fig. 4). Gradu-

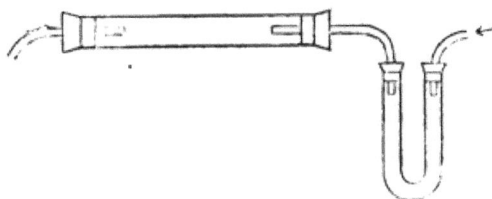

Fig. 4.

ally heat the glass tube containing the copper until it is red-hot. By means of a pair of bellows pass a *very slow stream* of air through the U tube and over the hot copper. The copper is soon changed to a black solid quite unlike the original red copper; this solid is *oxide of copper.* Continue heating until the change seems to be complete.

Allow the tube and its contents to cool, and then counterpoise again; there has been an increase in weight.

Inasmuch as the copper, iron, and magnesium, have all been heated in air, and the new matter produced has in each case weighed more than the matter before heating, we may conclude, provisionally, that the three chemical changes are analogous; and that, probably, their causes are similar. In each case there has been a combination with the heated substance of some other kind, or other kinds, of matter. The most likely source of this other kind of matter, considering the conditions of the experiments, is the air.

We must now make two assumptions, which can be, and have been, proved by accurate experiments. We shall assume (1) that the air is a mixture of at least two gases called oxygen and nitrogen; (2) that water is a compound of two gases, hydrogen and oxygen. If then hydrogen is brought into contact with a heated solid substance and water is produced, it follows that oxygen must have been taken away from the heated solid by the hydrogen.

Now remove the bellows, and in their place put a flask containing zinc and a little dilute sulphuric acid (Fig. 5). By

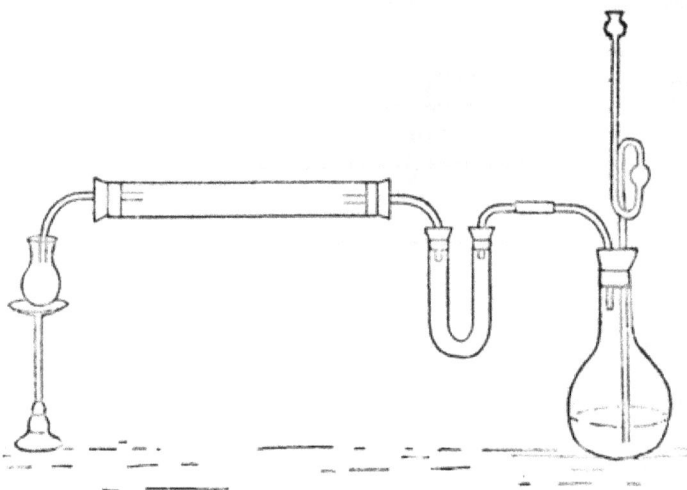

Fig. 5.

the interaction of these hydrogen is produced. Keep the flask in cold water. Let the hydrogen pass through the U tube containing calcium chloride, and then through the tube containing the oxide of copper. Let the end of this tube pass into a small *dry* flask. When the hydrogen has been passing for quite 5 minutes, *but not before,* begin gradually to heat the tube.

*The flask is dried by rinsing it out with water, then with a little alcohol, allowing the alcohol to drain out of the flask, warming the flask in the flame of a Bunsen-lamp, blowing air into it by a bellows, and repeating the warming and blowing in air two or three times.*

After a little you notice that red copper is produced in the tube, and that drops of water are formed in the small flask. Continue the passage of the hydrogen over the hot copper oxide so long as any water seems to be produced, or any change to occur in the appearance of the matter in the tube. Towards the close of the experiment heat the tube strongly, to make sure that no water remains in the tube but that all is

driven over into the small flask.    Then allow the tube to cool,
and counterpoise again.

The weight is the same as it was at the beginning of this
series of experiments.

You have therefore proved, on the basis of certain assumptions, that when copper is heated in air it combines with
oxygen in the air to produce a new kind of matter called
copper oxide ; and that the weight of the copper oxide thus
produced is greater than that of the copper from which it has
been produced.

By experiments too difficult to be performed at present it
can be proved that the difference between these weights is the
weight of the oxygen which has combined with the copper.

The change which occurs when magnesium, or iron, is burnt,
also consists in combination of the magnesium, or iron, with
oxygen in the air.

**Exp. 4.**    Powder some crystals of *potassium chlorate* in a
dry mortar ;    dry the powder by pressing, not rubbing, it
between filter paper, and place a little of it in a *dry* test tube.

Counterpoise the tube with its contents ;    heat gently until
the potassium chlorate melts and evolves a gas, then raise the
temperature a little.    Prove that the gas which is evolved is
not air, by bringing into the tube a chip of wood which is just
glowing ;    the wood bursts into flame.    The gas evolved is
*oxygen.*    Continue heating for a little, then allow the tube to
cool, and counterpoise again.

The contents of the tube weigh less than the potassium
chlorate did.    But if this is so, it is probable, although not
yet experimentally proved, that the quantity of gas which was
evolved also weighs less than the potassium chlorate.

Prove that the solid residue in the tube is not potassium
chlorate, by dissolving it, and a little potassium chlorate, separately, in water, and adding to each solution a few drops of
solution of silver nitrate; in the case of the chlorate no visible
change occurs, in the other case a white pp. is produced (this
pp. is *silver chloride* : the solid obtained by heating potassium
chlorate is called *potassium chloride*).

The substance potassium chlorate has been changed, by
the action of heat, into at least two different kinds of matter,
*oxygen* and *potassium chloride;* the mass of each of these is less
than the mass of the potassium chlorate changed.

**Exp. 5.**    Place a little dry *cuprous oxide* in a dry tube of glass; counterpoise; heat; notice the change which occurs; cool; counterpoise again. A new kind of matter different from, and weighing more than, the original cuprous oxide, has been produced. This new kind of matter is called *cupric oxide*.

In Exps. 1, 2, and 3 you changed specified masses of certain kinds of matter each into another kind of matter; the mass of the new kind of matter was in each case greater than the mass of the original kind of matter. The changes which *magnesium*, *iron*, and *copper* underwent are representative of the kind of chemical changes which Elements undergo. An Element is chemically changed by adding to, or combining with, it some kind, or kinds, of matter different from itself; the product, or the sum of the products if there are more than one, of such a change weighs more than the element weighed.

In Exp. 4 you changed a specified mass of potassium chlorate into two different kinds of matter, each unlike, and each weighing less than, the original matter. But in Exp. 5, you changed a specified mass of a certain kind of matter, cuprous oxide, into another kind of matter different from, and weighing more than, itself. The change which *potassium chlorate* underwent, and also the change which *cuprous oxide* underwent, are representative of the chemical changes which Not-elements undergo. A Not-element is chemically changed, either, as an element is, by adding to, or combining with, it some other kind or other kinds of matter, or by separating it into two or more different kinds of matter each unlike the other and unlike the original substance, and each weighing less than the original substance did before the change.

Exp. 6. Carefully counterpoise a small piece of *tin-foil* in a crucible; place the crucible on a sand-bath in the draught cupboard; heat the sand gently; let a *little* nitric acid fall on to the tin *drop by drop*. When the contents of the crucible have turned to a white powdery solid cease to add acid, but continue heating, and raising the temperature, until fumes of acid are no longer given off. Then cool, and counterpoise; the solid in the crucible is different from the tin used and it weighs more than the tin.

Reasoning solely on the results of this experiment and of Exps. 1 to 5, would you be inclined to place tin in the class of *elements* or in that of *not-elements*?

Do you think that tin can be placed in one or other of these classes solely on the evidence given by this experiment?

*Reference to* " ELEMENTARY CHEMISTRY." Chap. II.

# CHAPTER III.

**Exp. 1.** You are given some finely powdered *sulphur* and some very finely divided *iron.* Each of these is an element.

(*a*) Place a *little* of each in a test tube, pour some carbon disulphide into each tube, shake up, and gently warm by placing the tubes in hot water. The sulphur slowly dissolves; the iron is unchanged. To prove that sulphur has dissolved, filter, and evaporate off the carbon disulphide by placing the liquid in a watch-glass on a water-bath.

*Never warm carbon disulphide over a flame; a mixture of air with carbon disulphide vapour is very explosive.*

(*b*) Shake up a little of the iron and the sulphur, separately, with water; the iron quickly sinks to the bottom of the tube, part of the sulphur floats in and on the surface of the water.

(*c*) Place a little of the iron and sulphur, separately, on clean sheets of paper; bring a magnet under each, touching the under side of the paper; send a stream of air from the mouth over the surface of the iron and the sulphur. The sulphur is quickly blown away, but the iron is more or less firmly held by the magnet.

Now make a mixture of the iron and sulphur in the ratio of 1 part of sulphur to 1¾ parts of iron, by weight. Examine this mixture by (*a*) placing a *little* of it in warm carbon disulphide; (*b*) shaking a little with water; (*c*) placing a little on paper, bringing a magnet under the paper, and blowing a rapid air-stream over the substance. Each substance in the mixture behaves in the same way as it did when unmixed with the other substance.

Place a *little* of the mixture of iron and sulphur in a clean *dry* test-tube, and heat by a Bunsen-lamp until the whole mass glows strongly; allow to cool; break the tube in a mortar, and pick out the broken glass. Examine the black solid thus formed by action on it of (a) carbon disulphide, (b) water, (c) a magnet. The substance appears to be homogeneous; it is not separated into unlike parts by any of the methods by which the mixture of iron and sulphur was separated into unlike parts.

A Compound of iron and sulphur, *iron sulphide*, is formed by heating these two elements together in the ratio of 1 part by weight of sulphur to $1\frac{3}{4}$ parts of iron. The properties of the compound are very different from those of either of the elements which have combined to form it. The compound cannot be separated into unlike parts by methods which succeed in separating a Mixture of the constituents of the compound, i.e. a mixture of iron and sulphur, into unlike parts.

**Exp. 2.** Place a little black *copper oxide* in a tube with some water; warm; no visible change occurs. Separate the liquid from the solid by filtration, and evaporate the filtrate to dryness; no solid remains; therefore none of the copper oxide dissolved in the warm water.

Place a little yellow *potassium chromate* in water, and warm; the yellow solid dissolves, forming a yellow coloured liquid. Evaporate this to dryness; the original yellow potassium chromate remains.

Make a mixture of copper oxide and potassium chromate by pounding the two substances together in a mortar. The mixture is yellowish-black. Examine it by a magnifying glass; you can distinguish black particles (copper oxide), and yellow particles (potassium chromate). Place some of the mixture in water; warm; filter; add more water to the remaining solid, again warm, and filter. Dry the black solid which remains by placing it in a steam-bath. Evaporate the yellow filtrate to dryness. Mix the yellow potassium chromate thus obtained with the black copper oxide also obtained from the mixture; a yellowish black substance, similar to the original mixture, is produced.

You have thus separated a Mixture into its constituent parts by making use of a certain physical property of each constituent; viz. solubility in water of one constituent, and insolubility of the other.

**Exp. 3.** Examine the physical properties—colour, appearance, hardness, &c.—of a piece of *copper*. Examine the physical properties—appearance, liquidity, &c.—of *sulphuric acid*. The two substances are most distinctly marked off from each other by their physical properties.

Cut the copper into very small pieces, place these in the sulphuric acid, and warm until a part of the copper has dissolved in the acid; pour off the blue coloured liquid into a basin; evaporate the liquid, in the draught cupboard, *nearly but not quite to dryness*; allow to cool. A blue, crystalline, solid is obtained, very unlike either the copper or the sulphuric acid by the interaction of which it has been produced. This blue solid is called *copper sulphate*.

Attempt to separate the copper sulphate into its constituents by (*a*) acting on it with water, it dissolves but is obtained unchanged on evaporating off the water; (*b*) acting on it with a mixture of alcohol and water, it dissolves but is obtained unchanged when the solvent is removed by evaporation; (*c*) acting on it with strong alcohol, it remains unchanged.

Now dissolve in water some of the copper sulphate which you have prepared, add a little sulphuric acid, and immerse in the liquid two platinum plates, each of which is connected with a galvanic battery (Fig. 6). The electric current passes

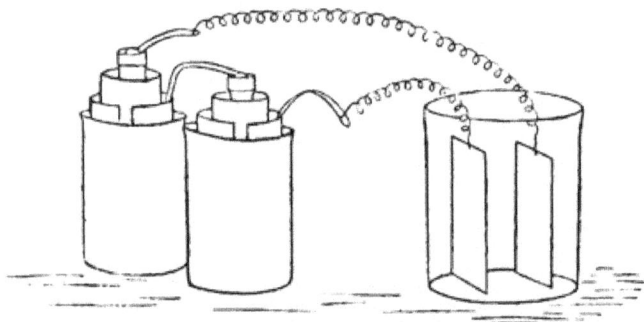

Fig. 6.

from one platinum plate to the other through the solution of copper sulphate. After a short time you notice that the platinum plate connected with the zinc plate of the battery is covered with a reddish solid. Allow the current to pass for

a little time; then remove the platinum plate and examine
the red solid deposited on it ; so far as you can judge, this
solid is copper.   That the solid is copper can be proved without doubt; but at present you must be content with such a
rough proof as is afforded by comparing, by means of the
senses, the red solid you have obtained with the copper given
you at the beginning of the experiment.

In this experiment you produced a Compound of copper
with sulphuric acid ; you failed to separate this into unlike
parts by methods which had already succeeded in separating
one or two Mixtures into unlike parts; but you separated the
Compound into unlike parts, one of which was known to be a
constituent of the compound, by using the agency of an electric
current.

*Reference to* "ELEMENTARY CHEMISTRY."   Chap. III.

# CHAPTER IV.

## CONSERVATION OF MASS OF MATTER.

**Exp. 1.** Fill a large test tube with water, cover the mouth with the thumb, and invert the tube in a small light basin partly filled with water. The tube should now be quite full of water; air must not be allowed to get in while the tube is being placed in the water in the small basin. If a ring of thick glass is slipped over the test tube the tube will stand steadily in the basin (Fig. 7). Place the small piece of zinc* given you in the basin, and bring it under the mouth of the tube (s. Fig. 7). Pour a very little concentrated sulphuric acid into a very small beaker. Place the small basin with its contents, and the small beaker containing sulphuric acid, on the pan of a fairly good balance, and counterpoise the whole. Without removing the basin from the balance-pan, pour the sulphuric acid into it, and replace the little beaker on the pan of the balance. A chemical change proceeds between the dilute sulphuric acid and the zinc; a gas collects in the tube; this gas is *hydrogen*. From time to time, as the change proceeds, and again when the change is finished, allow the balance to swing; the mass of the matter in one pan remains equal to that of the counterpoise in the other.

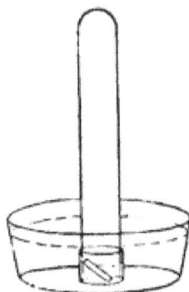

Fig. 7.

New kinds of matter have been produced in this experiment, but the sum of the masses of these is equal to the sum of the masses of the kinds of matter present before the change began.

* *Note to Demonstrator.* The piece of zinc should be such that when it dissolves in acid hydrogen is produced sufficient to fill the tube about ¾ths. ·08 grams zinc produce about 25—30 c.c. hydrogen.

**Exp. 2.** Collect *carbon dioxide* in a *perfectly dry* test tube (*s. Chap. I. Exp.* 6). When the tube is full, pour a little concentrated potash solution into it, and instantly cover the mouth of the tube with the thumb; shake briskly; invert the tube under a little water in a basin, and withdraw the thumb; the water rushes into and nearly (or perhaps quite) fills the tube. Carbon dioxide is therefore easily and rapidly dissolved by a concentrated aqueous solution of caustic potash.

You are now given an apparatus formed of two light test tubes and a small piece of fairly wide glass tubing (Fig. 8).

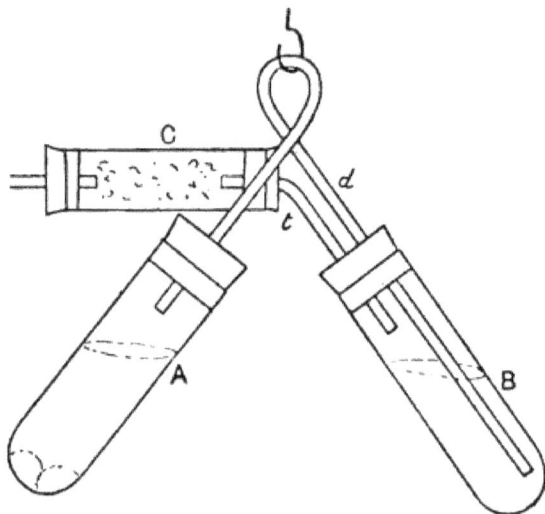

Fig. 8.

The tube *C* contains small pieces of solid caustic potash; fill *B* about ¾ths with a solution of caustic potash (1 part solid potash in 2 parts water, by weight). Place a couple of little bits of *marble* in *A*, each about 5 mm. (say ¼ in.) diameter; and pour in a *very dilute* solution of *hydrochloric acid* to about the height shewn in the figure. Insert the corks in *A* and *B*, but arrange the tube *d* so that it does not dip under the surface of the potash solution in *B*, and do not place *C* in connection with *B*; suspend the apparatus from the hook at the end of one arm of the balance, and place *C* on the pan of the balance; on the other pan place a counterpoise which is *very slightly* lighter than the whole apparatus, and allow the balance to swing freely. After a few minutes the position of the pointer of the balance indicates that the system of tubes has lost weight slightly ; the stream of carbon dioxide has swept the air out of the tubes, and some of the carbon dioxide has also passed out of the apparatus.    Now adjust the tube *d* as shewn in the figure, attach *B* to the small tube *t*, and arrange the counterpoise so that the pointer indicates that the balance is in equilibrium.    Allow the balance to swing, and observe the course of the change.    *Marble* (calcium carbonate) and *hydrochloric acid* are interacting to produce *water, carbon dioxide*, and *calcium chloride*, but everything is retained in the apparatus.    The total mass of the matter remains unchanged.

The total mass of matter taking part in a chemical or physical change is constant ; the sum of the masses of the different kinds of matter produced in a chemical change is equal to the sum of the masses of the different kinds of matter which by their interaction produced the new kinds of matter.

*Reference to* " ELEMENTARY CHEMISTRY."    Chap. IV.

# CHAPTER V.

A. *Chemical change is sometimes brought about by the agency of heat.*

**Exp. 1.** Heat a little *red oxide of mercury* in a dry test tube, keeping the mouth of the tube loosely covered with the thumb. A sublimate of mercury appears on the colder parts of the tube (*s. Chap. I. Exp.* 3); continue heating until the lower part of the tube is red hot; then remove the thumb and bring a glowing splint of wood into the tube; the glowing wood bursts into flame; *oxygen* is being evolved from the heated mercury oxide (*comp. Exp.* 4 *in Chap. II.*).

In *Chap. II. Exp.* 4, the compound potassium chlorate was changed by the action of heat into potassium chloride and oxygen. In *Chap. I. Exp.* 3, the compound lead nitrate was changed by the action of heat into nitrogen tetroxide and lead oxide. In *Chap. III. Exp.* 1, a mixture of iron and sulphur was changed by the action of heat into a compound of iron and sulphur (iron sulphide).

B. *Chemical change is sometimes brought about by the interaction of two or more different kinds of matter at ordinary temperatures.*

Compare *Exps.* 4, 5, 6 *in Chap. I.* and *Exp.* 1 *in Chap. IV.*

**Exp. 2.** To a solution in water of *barium nitrate* add a dilute solution of *sulphuric acid*; a white solid is at once produced (*barium sulphate*).

**Exp. 3.** To a solution of *silver nitrate* add a dilute solution of *hydrochloric acid*; a white solid forms at once (*silver chloride*).

C. *Chemical change is frequently brought about by the interaction of two or more kinds of matter aided by heat.*

Compare *Exp.* 1, *Chap. I. Exps.* 2, 3, 5, *and* 6, *Chap. II.* and *Exp.* 3, *Chap. III.*

**Exp. 4.** Place a little *sulphur* in the cup of a *deflagrating spoon* (Fig. 9), and bring it into a jar of oxygen; no change occurs. Now heat the sulphur in a Bunsen-lamp until it begins to burn and then plunge it into the jar of oxygen. The sulphur burns brilliantly; when the burning has ceased, withdraw the spoon, pour some distilled water into the jar and shake briskly. Taste a small drop of the solution; it is sour. Pour some blue litmus into the solution; the colour changes to red. The sulphur and oxygen have combined to form a new substance, *sulphurous oxide*, which has dissolved in the water.

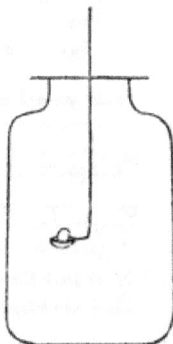

Fig. 9.

**Exp. 5.** Pour a little dilute solution of *nitric acid* over a piece of *copper* in a test tube; no visible change occurs. Warm the contents of the tube; chemical change soon begins.

How do you know that the change which occurs is chemical?

**Exp. 6.** Put a small piece of *charcoal* with some *potassium chlorate* in a dry test tube; no change occurs. Heat the contents of the tube until the potassium chlorate melts and gives off gas; the carbon burns brilliantly forming *carbon dioxide*. Prove that carbon dioxide is produced by allowing some of the gas to fall into a tube containing a little *lime water* (solution of lime in water), and shaking up; a white pp., *calcium carbonate*, is produced (*s.* Fig. 10).

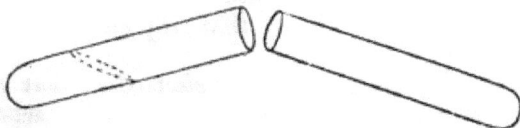

Fig. 10.

2—2

*The production of this white pp. in lime water is a test whereby the presence of the gas carbon dioxide may be detected.*

Carbon dioxide is a compound of carbon and oxygen: what was the source of the oxygen in this experiment?

In the preceding experiments it is to be noticed that when a chemical change occurred by the interaction of two or more substances, one at least of the substances was in the liquid or gaseous state. The change of a mixture of iron and sulphur to iron sulphide appears to be an exception; but when we remember that sulphur is easily melted, this change will be seen to be no exception to the statement.

D.   *Chemical change is sometimes brought about by the electric current.*

Compare *Exp. 3, Chap. III.*

**Exp. 7.**   Pass an electric current through a solution of *silver nitrate* to which a little nitric acid has been added. Silver separates at the negative *electrode.*

The arrangement shewn in Fig. 11 is a simple one for such

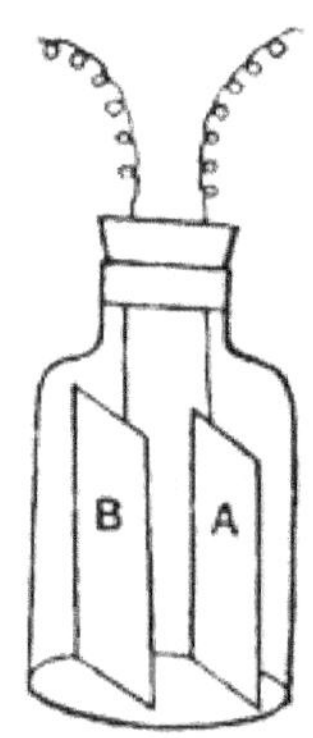

Fig. 11.

experiments as the electrolytic decomposition of silver nitrate solution. A and B are plates of platinum attached to platinum wires which pass through a cork in a small wide-mouthed bottle; the cork fits loosely so that any gas evolved during the electrolysis may escape.

The platinum plates are called *electrodes*; that on which the silver (or copper, *s. Exp. 3, Chap. III.*) separates is sometimes called the *kathode* and the other the *anode.*

**Exp. 8.** Pass an electric current through *water* containing
a little sulphuric acid.   Arrange the apparatus so that each
electrode (of platinum) passes a little way into a large tube
full of acidulated water inverted over it (Fig. 12).   Gas

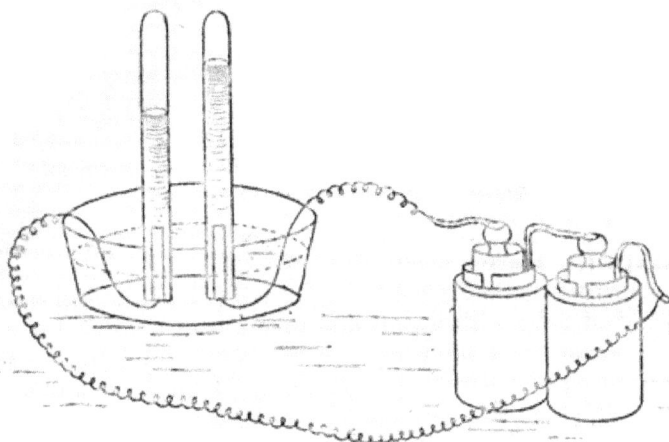

Fig. 12.

collects in each tube; after a time stop the current; cover the
mouth of the tube which contains the larger volume of gas with
the thumb, keeping the tube under the liquid all the time;
invert the tube, withdraw the thumb, and *at once* bring a
lighted taper to the mouth of the tube ; the gas takes fire with
a slight explosion, and burns with a nearly colourless flame.
Now get ready a glowing chip of wood ; cover the mouth of
the other tube with the thumb, invert the tube, withdraw the
thumb, and plunge the glowing chip into the gas ; the wood
bursts into flame, but the gas does not take fire.

The first gas is *hydrogen*, the second is *oxygen*.

*You have learned in this experiment how to detect hydrogen
and oxygen and to distinguish them from each other.*

# CHAPTER VI.

## CHEMICAL PROPERTIES OF WATER.

**Exp. 1.** Place a small piece of *sodium* in a little cage of wire gauze attached to a glass rod (Fig. 13). Fill a large test tube with water, and invert it in a small basin of water; hold the tube with one hand, and with the other bring the wire cage containing the sodium under the water so that the gas which at once begins to bubble through the water passes into the tube and collects there. When the tube is full of gas, cover the mouth with the thumb, invert the tube, and bring a lighted taper to the mouth; the gas takes fire and burns with a pale almost non-luminous flame. The gas is *hydrogen*.

Evaporate the water in the basin to dryness; the white solid which remains is a compound of sodium, hydrogen, and oxygen; it is called *sodium hydroxide*, or *caustic soda*. (The composition of this compound cannot be proved at present.)

Fig. 13.

By the interaction of sodium and water, hydrogen and a compound of sodium with hydrogen and oxygen, have been formed. Sodium is an element; if this is taken as proved, it follows that the hydrogen evolved as gas in the foregoing experiment, and also the hydrogen and oxygen which combined with the sodium, must have formed part of the water at the beginning of the experiment. (Here we assume that the material of the vessels was not chemically changed during the process.) Water therefore is a compound of hydrogen and oxygen.

In *Exp.* 8, *Chap. V.* water was decomposed into hydrogen and oxygen by the action of an electric current. In *Exp.* 3, *Chap. II.* water was produced by passing hydrogen over hot copper oxide.

Water interacts with certain elements at high temperatures to produce oxygen and a compound of hydrogen with the interacting element*.

The composition of water can be proved only by very accurate quantitative experiments. The composition of water is ;—

| *By weight.* | *By volume; water-gas.* |
|---|---|
| Hydrogen 1 part⎫ form 9 parts | Oxygen 1 vol.⎫form 2 vols. |
| Oxygen 8 parts⎭ of water. | Hydrogen 2 vols.⎭ of water-gas. |

(*s.* "ELEMENTARY CHEMISTRY." Chap. VII. pars. 92 to 96.)

Hydrogen and oxygen are not generally prepared by decomposing water.

**Exp. 2.** Fit up an apparatus like that represented in Fig. 14. Place some granulated *zinc* in the bottle, cover it

Fig. 14.

with water, and pour a little *dilute sulphuric acid* into the bottle through the funnel-tube. The zinc and sulphuric acid react to produce *zinc sulphate*, which remains dissolved in the water, and *hydrogen*, which passes off as gas. While the air

* The production of oxygen and hydrogen chloride by the interaction of chlorine and steam at a red heat should be demonstrated to the student.

is being swept out of the apparatus by the stream of hydrogen, fill two wide-mouthed bottles with water, and invert each in the water in the trough (as in fig. 15). Now fill a test tube with water and invert it over the end of the tube from which the hydrogen is escaping; when the tube is full of hydrogen bring it quickly mouth downwards to a flame; if the hydrogen ignites without explosion all the air has been removed from the apparatus; if an explosion attends the ignition of the hydrogen the tube contained a mixture of air and hydrogen, and therefore the whole of the air has not been swept out of the apparatus by the stream of hydrogen. When the hydrogen is free from air, fill the two bottles with it. Keep the bottles standing mouth downwards in the water until they are wanted. Set aside the flask containing the zinc and sulphuric acid (s. Exp. 8, p. 26).

*A mixture of air and hydrogen is very explosive: never bring a light near any apparatus containing hydrogen, nor collect hydrogen, until you have proved that there is no air mixed with the hydrogen.*

**Exp. 3.** Fit up a dry flask with a tightly fitting cork and delivery tube as shewn in Fig. 15. Place the mixture of

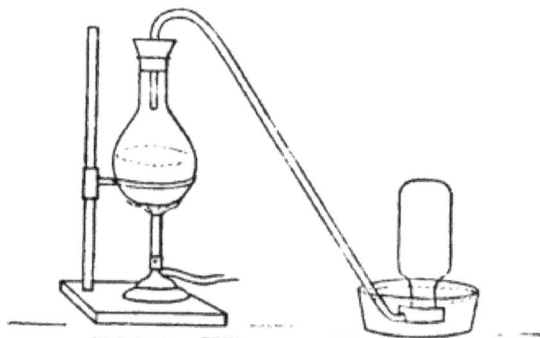

Fig. 15.

dry powdered *potassium chlorate* and *manganese dioxide* given you in the flask*; fill four bottles with water and have them

* *Note to Demonstrator.* About 1 part dry manganese dioxide is mixed with 3 parts dry powdered potassium chlorate. A *little* of the mixture should be tested by heating in a tube to see that oxygen comes off quietly. Manganese dioxide is sometimes adulterated with carbonaceous matter; the use of such adulterated material is *highly dangerous*. About 8 grams of the mixture may be used for each litre of oxygen required.

ready standing mouth downwards in the trough. Heat the flask, gently at first, then more strongly; after gas has been coming off for a little, collect a test tube full of the gas and bring a glowing chip of wood into it; if the wood at once bursts into flame you conclude that the gas coming off is approximately pure oxygen. Fill each of the four bottles with the oxygen, and keep them standing in the trough until they are wanted. Before removing the lamp from beneath the flask remove the delivery tube from the water. Set aside the flask containing the materials from which the oxygen was made (s. Exp. 8, p. 26).

**Exp. 4.** Light a taper; remove a bottle of *hydrogen* from the trough, hold it mouth downwards, and bring the taper to the mouth of the bottle; the hydrogen takes fire; move the taper into the bottle, the taper is extinguished.

Remove a bottle of *oxygen* from the trough and bring the lighted taper to the mouth of the bottle, the oxygen does not take fire; move the taper into the bottle, the taper burns brilliantly.

Hydrogen under ordinary conditions is a *combustible gas*; oxygen under ordinary conditions is a *supporter of combustion*. These terms are purely relative; under certain conditions oxygen is combustible and hydrogen is a supporter of combustion.

**Exp. 5.** Place a bottle of *hydrogen* and one of *oxygen*, at some little distance apart, mouth upwards, on the table; after one or two minutes bring a lighted taper into each bottle; the visible phenomena shew that the bottle which contained hydrogen now contains only air, but that the other bottle still contains oxygen.

Hydrogen is therefore much lighter than air; oxygen is either heavier than air, or the specific gravities of oxygen and air are nearly equal.

**Exp. 6.** Arrange a piece of porous wood *charcoal* on the end of a wire which passes through a piece of stout card-board sufficiently large to cover the mouth of the third bottle of oxygen. Ignite the charcoal; reverse the bottle of *oxygen*, and plunge the burning charcoal into it (Fig. 16). The charcoal burns brilliantly and rapidly in the oxygen. When the burning

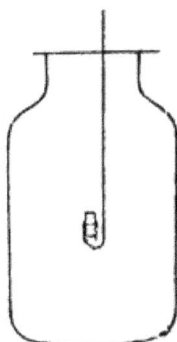

Fig. 16.

is finished pour some distilled water into the bottle; shake briskly; pour a little of this solution into a test tube, and prove the presence of *carbon dioxide* in it by the lime water test (*Chap. V. Exp.* 6).

To the remainder of the solution of carbon dioxide in the bottle add some blue litmus solution; the colour of the litmus changes to purplish-red.

**Exp. 7.** Place a small dry piece of *sodium* in a *clean* deflagrating spoon; heat the sodium until it begins to burn and then plunge it into the remaining bottle of oxygen, taking care that the spoon does not touch the bottle. The sodium burns brilliantly and rapidly. When the combustion is finished, pour some distilled water into the bottle over the spoon; shake briskly, and then add a little litmus solution; the blue colour is deepened in tint: now pour in a little red litmus solution; the colour is changed to blue.

In Exps. 6 and 7 oxides were formed; oxide of carbon in Exp. 6, oxide of sodium in Exp. 7: both oxides dissolved in water; one of these solutions turned blue litmus red, the other turned red litmus blue. We shall find as we proceed that many oxides dissolve in water to form solutions which affect the colour of litmus as carbon dioxide does, and that many form solutions in water which affect the colour of litmus as sodium oxide does. Moreover we shall find that all the oxides in each of these classes have several properties in common besides that of changing the colour of litmus. (*s. Chap. VII.*)

**Exp. 8.** Turn back to the flasks containing the *materials from which hydrogen and oxygen were obtained.* Pour a portion of the contents of the flask in which hydrogen was generated through a filter, and set the filtrate evaporating in the draught-place, on a water bath. Add some distilled water to a portion of the contents of the other flask, warm for some time, and filter; set the filtrate evaporating. Repeat the treatment with warm water as long as the water continues to dissolve out anything, that is as long as a few drops of the filtrate leave a solid residue when evaporated to dryness in a watch-glass. Dry the insoluble residue and compare it with the manganese dioxide given you; as far as you can judge they are identical. This conclusion, founded on the unconfirmed evidence of the senses and therefore by no means to be trusted, is confirmed by accurate quantitative analysis.

Now turn to the two filtrates; each is probably evaporated to dryness. The liquid obtained by adding warm water to the solid matter which remained from the preparation of oxygen has disappeared, and in its place you find a white solid. Dissolve a little of this solid in water, and add a few drops of silver nitrate to the solution; a white pp. forms; add a little nitric acid, the white pp. remains.

*The production of a white pp. (silver chloride), insoluble in dilute nitric acid, when silver nitrate is added to a liquid, is a test for the presence of a chloride in that liquid.*

You conclude therefore that when a mixture of potassium chlorate and manganese dioxide is heated, there are produced oxygen, a chloride (probably potassium chloride), and manganese dioxide. This conclusion has been confirmed by accurate quantitative experiments.

The liquid filtered from the unchanged zinc remaining in the flask used for preparing hydrogen has evaporated and left a white solid. Heat this solid on the water bath so long as an acid-smelling vapour is given off; then dissolve the residue in a little warm water, filter if necessary, evaporate the filtrate until a drop taken out by a glass rod and placed in a watch-glass partly solidifies on cooling, and then allow to cool; white crystals form in the liquid; pour off the liquid, remove the crystals by the help of a spatula to filter paper, and press them until nearly dry; again dissolve the crystals in as little boiling water as possible, evaporate slightly, and allow to cool; remove the crystals which form to filter paper and dry them by pressure. By these processes of pressing between porous paper, dissolving in water, and crystallising, you obtain the crystals free from adhering sulphuric acid. To make sure that all sulphuric acid has been removed, dissolve a little of the last crop of crystals in water and add a drop of blue litmus solution; if the colour of the litmus is unchanged all sulphuric acid has been removed. If however the litmus turns red the whole of the crystals must be again recrystallised and dried by pressure between porous paper. When you have obtained the crystals free from adhering acid, dissolve them in water and divide the solution into two parts. To one part add barium nitrate (or barium chloride) solution; a white pp. forms; add a little hydrochloric acid, the pp. remains unchanged. The white pp. is barium sulphate; its production, and insolubility in hydrochloric acid, proves that the solution contained a sulphate. To the other part of the solution of the crystals add caustic potash solution drop by drop; a white gelatinous pp. forms; add more potash, the pp. dissolves. The white pp. is zinc hydroxide ; the production of this pp., and its solubility in much caustic potash, proves the presence of a compound of zinc in the original liquid.

*The production of a white pp. (barium sulphate), insoluble in hydrochloric acid, when barium nitrate solution is added to a liquid, is a test for the presence of a sulphate in that liquid.*

*The production of a white gelatinous pp. (zinc hydroxide), soluble in much caustic potash, when a little caustic potash solution is added to a liquid, is a test for the presence of a compound of zinc in that liquid.*

You conclude that when zinc and dilute sulphuric acid interact, hydrogen and zinc sulphate are produced. This conclusion has been verified by accurate quantitative experiments.

From the experiments in this chapter you have learned, (1) how to prepare hydrogen and oxygen; (2) something regarding the chemical changes which proceed when hydrogen is prepared (a) by the interaction of zinc and dilute sulphuric acid, (b) by the interaction of water and sodium, and when oxygen is prepared by the action of heat on potassium chlorate mixed with manganese dioxide; (3) that, under ordinary conditions, hydrogen is combustible and oxygen is a supporter of combustion ; (4) that hydrogen is much lighter than air, but that equal volumes of oxygen and air are nearly the same weight ; (5) that a mixture of air and hydrogen explodes when a flame is brought near it ; (6) that carbon burns in oxygen to form an oxide, an aqueous solution of which turns blue litmus red ; (7) that sodium burns in oxygen to form an oxide, an aqueous solution of which turns red litmus blue. You have also learned incidentally how to detect (a) a chloride, (b) a sulphate, (c) a compound of zinc, in solutions ; and how to purify a solid soluble in water from other substances more soluble than itself.

Water is a compound of hydrogen and oxygen; let us examine a few of its properties.

**Exp. 9.** To three separate portions of distilled *water* in basins add, (1) some powdered *copper sulphate crystals*, (2) powdered *potassium nitrate*, (3) powdered *tartar-emetic* ; warm each basin slightly, and if the solids do not wholly dissolve add a little more water. After a time the whole of each solid has disappeared in the water. Evaporate a portion of each solution to dryness; so far as you can judge by their appearances, the solids obtained are the same as those originally dissolved in the water.  Accurate experiment confirms this conclusion.

In these three cases water has acted as a solvent.  Water dissolves very many chemical compounds, and a few chemical elements, without changing their composition.

**Exp. 10.**  To two separate portions of distilled *water* in basins add (1) *anhydrous copper sulphate*, (2) solid *sulphur trioxide**; each dissolves rapidly, the latter with a hissing sound.  Evaporate the solutions somewhat, and allow to cool; blue crystals are obtained from the first, and a thickish, very acid, corrosive, liquid from the second.  Both products are different from the substances added to the water.  In one case *hydrated copper sulphate*, in the other case *sulphuric acid*, has been formed.

In these cases water has acted as a solvent, but at the same time it has interacted with the substance added to it and has produced a new substance.

*Reference to* " ELEMENTARY CHEMISTRY."  Chap. VII. pars. 90 to 106.

---

\* This should be supplied in small sealed tubes; it is easily prepared by warming Nordhausen sulphuric acid and leading the vapour into a series of dry test tubes, each of which is sealed off when it contains a small quantity of the trioxide.

# CHAPTER VII.

## CLASSIFICATION OF OXIDES.

**Exp. 1.** To three separate quantities of *water* in basins add (1) *phosphorus pentoxide*, (2) *sulphur trioxide*, (3) *sodium oxide*\*. When the oxides are dissolved examine the solutions as follows :—

(*a*) to two small quantities of each add a drop of blue litmus, and a drop of red litmus, respectively ;

(*b*) in a small quantity of each solution place a little piece of the metal zinc ;

(*c*) to a small quantity of each solution add a little sodium carbonate, and notice whether carbon dioxide is given off.

The solutions of phosphorus pentoxide and sulphur trioxide turn the blue litmus red, dissolve the zinc more or less rapidly, and dissolve the sodium carbonate with evolution of carbon dioxide.

The solution of sodium oxide turns the red litmus blue, does not dissolve the zinc, at least not so long as the solution remains cold, and does not evolve carbon dioxide from the sodium carbonate.

**Exp. 2.** Boil the remainder of the solution of *phosphorus pentoxide* for some time, and then *very cautiously and slowly* pour into it the solution of *sodium oxide* until the liquid is *as nearly neutral as you can make it*. To determine when this point is reached, arrange several small pieces of red and blue litmus paper on a piece of clean paper on a porcelain slab ; stir the solution of phosphorus pentoxide with a clean rod as you slowly pour in the solution of sodium oxide ; from time to

* *Caustic soda* may be used in place of the oxide.

time remove a drop of the liquid and let it fall on to one
of the pieces of blue litmus paper; when a drop of the liquid
produces only a slight reddening of the paper, dilute the
sodium oxide solution considerably (by pouring water into
it) and add the dilute solution one drop at a time.
Proceed thus until a drop of the liquid scarcely affects the
colour either of blue or red litmus paper. Should you add
too much sodium oxide solution (i.e. should a drop of the
liquid in the basin turn red litmus blue) you must dissolve a
little more phosphorus pentoxide in hot water and add this
drop by drop, to the liquid until a drop of it scarcely turns
red litmus blue or blue litmus red.

Place the *neutral solution* over a *low* flame to evaporate.
Meanwhile *neutralise* the solution of *sulphur trioxide* remaining from Exp. 2, by solution of *sodium oxide*, in the way
already described. Place the *neutral solution* over a *low*
flame to evaporate.

When both solutions have evaporated to dryness there
remains in each case a white solid. Remove a little of these
solids by means of clean spatulas to test tubes, dissolve in
water, divide each into two parts, and to one part add a
drop or two of blue litmus, and to the other part a drop
or two of red litmus; the solutions are almost or quite *neutral
to litmus.*

Dissolve in water a small portion of the white solid obtained
by neutralising the solution of phosphorus pentoxide by soda;
add a little nitric acid, and then heat to boiling; add a good
deal of ammonium molybdate solution, and remove from the
source of heat; a yellow pp. forms slowly.

*The formation of a yellow pp. by adding ammonium molyb-
date to a hot solution containing nitric acid is a test for the
presence of a phosphate in the solution.* [The yellow pp. is a
compound of molybdenum trioxide with ammonium phosphate.]

Dissolve in water a little of the white solid obtained
by neutralising the solution of sulphur trioxide by solution
of sodium oxide, and test for a sulphate.

Take a little piece of *clean* platinum wire; hold it in the
flame of a Bunsen-lamp, near the outer edge towards the
lower part of the flame; if the wire is quite clean no colour is
given to the upper part of the flame; if the flame is coloured,
the wire must be dipped in concentrated hydrochloric acid,
held in the flame, again dipped in the acid, and again held
in the flame, until it ceases to give any colour to the upper part

of the flame. Now moisten the wire, take up on it, successively, a very small particle of each of the solids you have prepared, and bring each into the lower part, near the outer edge, of the flame of a Bunsen-lamp; the upper part of the flame is at once coloured yellow. This colour proves the presence of a *compound of sodium.*

*All compounds of sodium impart a yellow colour to a non-luminous flame in which they are volatilised; the colour is scarcely visible through a blue glass.*

From these experimental results you conclude, (1) that when a solution in hot water of phosphorus pentoxide is neutralised by an aqueous solution of sodium oxide there is probably produced a *phosphate of sodium*; (2) that when a solution in water of sulphur trioxide is neutralised by an aqueous solution of sodium oxide there is probably produced a *sulphate of sodium.* These compounds are *salts.*

To prove these conclusions quite satisfactorily, it would be necessary to make accurate quantitative experiments.

Oxides which resemble phosphorus pentoxide and sulphur trioxide are called *acidic oxides.* These oxides interact with water to form acids; or they are obtained from acids by removing water from them.

Oxides which resemble sodium oxide are called *basic oxides.* These oxides interact with acids to form salts and water; some of them dissolve in water to form *alkalis.*

**Exp. 3.** *Zinc oxide is a basic oxide.* Place some dilute *sulphuric acid* in a basin; warm the acid, and add *zinc oxide*, little by little, until some of the oxide remains undissolved (i.e. add an *excess* of zinc oxide); now pour the liquid through a filter into another basin, and evaporate the filtrate *nearly* to dryness over a low flame. Allow to cool, collect the solid which has formed, and free it from all adhering *mother liquor* in the manner described in *Exp. 8, Chap. VI.*

*When crystals separate from a liquid, the remaining liquid is called the mother liquor.*

Finally dissolve the crystals you obtain in water and test the solution for (1) zinc, (2) a sulphate (*s. Exp. 8. Chap. VI.*). The crystals are *zinc sulphate.* But zinc sulphate is a salt; therefore zinc oxide is a basic oxide as it has reacted with sulphuric acid to form a salt.

**Exp. 4.** You are given an aqueous solution of *nitrogen pentoxide* and are asked to prove that this is an *acidic oxide.*

You are also given a solution of *potassium oxide* which you know to be a *basic oxide*.

Neutralise the solution of nitrogen pentoxide by the solution of potassium oxide; evaporate *nearly* to dryness; collect the crystals which form on cooling, and purify them by recrystallisation and pressure between porous paper. Prove that the solid thus obtained is *potassium nitrate* by (1) proving it to be a potassium compound by the flame test (*s. Exp.* 2 *of this Chap.*); (2) proving it to be a nitrate.

*Potassium compounds give a pale lavender colour to a non-luminous flame in which they are volatilised.*

*To test for a nitrate, dissolve in water, add a good deal of concentrated sulphuric acid, mix well by pouring from one tube to another, cool completely, hold the tube in a slanting direction, and pour into it very slowly a solution of ferrous sulphate prepared in cold water. The production of a brown-black colour in the liquid shews the presence of a nitrate.*

[The brown-black colour is the colour of a compound of nitric oxide—a product of the interaction of the sulphuric acid and the nitrate—with ferrous sulphate; this compound remains in solution; it is very easily decomposed by heat; if water or an aqueous solution mixes freely with fairly concentrated sulphuric acid heat is produced; hence the reason for conducting the test as directed.]

Never add concentrated sulphuric acid to a warm liquid, else most disastrous results may follow.

**Exp. 5.** You are asked to prove that *barium oxide* is a *basic oxide*. To a moderate quantity of dilute warm sulphuric acid in a basin add barium oxide; the oxide appears to remain unchanged; it really interacts with the acid to produce barium sulphate, but this salt is insoluble in water and dilute sulphuric acid. Continue to add the barium oxide to the acid until the liquid shews only a slight reddening action on litmus paper; add one or two drops of dilute sulphuric acid; warm for a few minutes, allow the solid matter in the basin to settle; pour off the liquid; add warm water; shake up briskly; allow to settle; pour off the liquid; repeat this process until the water ceases to redden blue litmus paper, i.e. until every trace of sulphuric acid is removed.

*This method of washing is called washing by decantation.*

You have now to prove that the white solid which remains in the basin is *barium sulphate*. To do this you must prove (1) that it is a compound of barium; (2) that it is a sulphate.

Bring a very little of the solid supported on a *clean* platinum wire (*s. Exp. 2 of this Chap.*) into the flame of a Bunsen-lamp : the flame is coloured pale grass-green.

*The production of a pale grass-green colour in a flame which before was non-luminous is a test for the presence of a compound of barium.*

When barium sulphate is heated for some time in contact with a large quantity of a concentrated solution of sodium carbonate, a part of the barium sulphate is changed to barium carbonate, and sodium sulphate is simultaneously formed; sodium sulphate dissolves easily in water.   On these reactions we may found a method for proving that the solid substance obtained as described is a sulphate.   Add sodium carbonate to some warm water as long as the solid dissolves ; put a little of this solution in a test tube ; to the rest of it add some of the solid which is to be tested, and allow the liquid with the solid suspended in it to boil for 5 minutes or so.   Meanwhile prove that the solution of sodium carbonate you have prepared is quite free from sulphate, by adding excess of hydrochloric acid and barium nitrate (*s. Exp. 8, Chap. VI.*).   Now return to the boiling solution of sodium carbonate holding the solid in suspension ; allow it to settle until you obtain a clear supernatant liquid ; pour off some of this (through a filter if necessary), and prove the presence of a sulphate in this liquid.

Barium oxide therefore interacts with sulphuric acid to form barium sulphate; but barium sulphate is a salt; therefore barium oxide is a basic oxide.

In this chapter you have learned something of the meanings of the terms *basic oxide* and *acidic oxide*; but you have seen that to understand the terms more fully it is necessary to know what is meant by an *acid*, and what is meant by a *salt*.

Incidentally you have learned how to detect *a nitrate* and *a phosphate*, and also compounds of *potassium*, of *barium*, and of *sodium*.   Finally you have learned how to *neutralise an acid by an alkali*.

*Reference to* "ELEMENTARY CHEMISTRY." Chap. VIII. pars. 123 to 126, and 130 to 132.

# CHAPTER VIII.

**Exp. 1.** To three quantities of dilute *sulphuric acid* in beakers * add, respectively, *zinc, zinc oxide,* and *zinc carbonate* ; gas is evolved in the first and third beakers ; no gas is evolved in the second beaker †.

Prove that the gas evolved during the interaction of the zinc and the acid is *hydrogen* ; and that the gas evolved in the interaction of the zinc carbonate and the acid is *carbon dioxide* (*s. Exp. 6, Chap. V.*). Warm the three liquids, and allow the reactions to proceed until all visible change has ceased ; now add a little more zinc, zinc oxide, and zinc carbonate, respectively ; warm again for a minute or two, and filter into basins. Evaporate the three filtrates to dryness over low flames ; collect the three solids, press them between porous paper, and crystallise each from water two or three times (*s. Exp. 8, Chap. VI.*). Prove that each solid (1) is a compound of zinc, (2) is a sulphate.

The solids have the same composition ; each is *zinc sulphate.* Note that the results of your experiments do not prove this conclusively ; accurate quantitative analysis is required. But we shall assume that such analyses have been made ; the result is that each solid has the same composition, and that each is zinc sulphate.

When zinc reacts with dilute sulphuric acid, zinc sulphate and hydrogen are formed ; when zinc oxide reacts with dilute

---

* About 100 c.c. of a mixture of 1 part strong acid with 8—10 parts water may be used.

† If the zinc oxide contains a little carbonate, as most specimens of the oxide do, a little carbon dioxide will be evolved.

3—2

sulphuric acid, zinc sulphate and water are formed (your experiments do not indicate or even suggest the production of water); when zinc carbonate reacts with dilute sulphuric acid, zinc sulphate, carbon dioxide, and water are formed.

[Note carefully what your own experiments have actually proved, and what further data would be required to prove these statements.]

**Exp. 2.** Perform an experiment similar to No. 1, but use a solution of *hydrochloric acid* in place of sulphuric acid, and *magnesium, magnesium oxide,* and *magnesium carbonate,* respectively, in place of zinc, zinc oxide, and zinc carbonate, respectively.

Prove that *hydrogen* is evolved in the reaction with magnesium; *carbon dioxide* in the reaction with magnesium carbonate; and no gas in the reaction with magnesium oxide.

Prove that the white solid obtained by evaporating each solution *just* to dryness at 100° is (1) a chloride, and (2) a compound of magnesium.

*The production of a white pp.* (magnesium-ammonium phosphate) *on adding sodium phosphate solution to a solution to which a large quantity of ammonium chloride and then excess of ammonia have been previously added, shews the presence of magnesium.*

Assuming—and this has been rigorously proved by quantitative analysis—that each of the solids obtained in this experiment is *magnesium chloride,* it follows that the reactions between hydrochloric acid and magnesium, magnesium oxide, and magnesium carbonate, are similar to the reactions between sulphuric acid and zinc, zinc oxide, and zinc carbonate.

These experiments enable us to gain a notion of the characteristic property of acids, viz. that they interact with metals, metallic oxides, and metallic carbonates, to form salts.

Zinc and magnesium are metals; zinc sulphate and magnesium chloride are salts; these salts were produced in your experiments by the interaction of sulphuric and hydrochloric acid, respectively, with a metal, a metallic oxide, and a metallic carbonate.

To understand more completely what is meant by saying that salts are formed by the interaction of metals, &c. with acids, it would be necessary to determine the compositions of the acids used, and of the salts obtained. Such determinations have been made. The results of the quantitative examination of the reactions used in Exps. 1 and 2 may be represented in chemical equations as follows * :—

*Reactions between zinc &c. and sulphuric acid.*

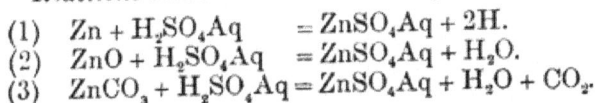

(1)  $Zn + H_2SO_4Aq$    $= ZnSO_4Aq + 2H.$
(2)  $ZnO + H_2SO_4Aq$    $= ZnSO_4Aq + H_2O.$
(3)  $ZnCO_3 + H_2SO_4Aq = ZnSO_4Aq + H_2O + CO_2.$

*Reactions between magnesium &c. and hydrochloric acid.*

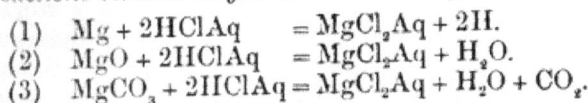

(1)  $Mg + 2HClAq$    $= MgCl_2Aq + 2H.$
(2)  $MgO + 2HClAq$    $= MgCl_2Aq + H_2O.$
(3)  $MgCO_3 + 2HClAq = MgCl_2Aq + H_2O + CO_2.$

A salt may at present be regarded, then, as a compound of a metal with the elements of an acid except hydrogen.

In the last chapter you learned that an acid reacts with an alkali to form a salt; alkalis are compounds of certain metals with oxygen and hydrogen; they are hydroxides of certain metals (*s. Chaps. VII. and X.*).

If an acid reacts with a metal, the oxide, and the carbonate, of a metal, to form a salt; if an acid also reacts with an alkali to form a salt; and if an alkali is the hydroxide of a certain class of metals, it is probable that an acid will react with any metallic hydroxide, whether it be an alkali or not, to form a salt. Let us put this hypothesis to the test of experiment.

**Exp. 3.**    Add (1) *excess of barium hydroxide* to *nitric acid,* and (2) *excess of ferric hydroxide* to *sulphuric acid,* and proceed as described in Exp. 1. After purifying the salts obtained prove that one is (1) a compound of barium (*s. Exp.* 5, *Chap. VII.*); (2) a nitrate (*s. Exp.* 4, *Chap. VII.*); and that the other is (1) a compound of iron, (2) a sulphate (*s. Exp.* 8, *Chap. VI.*).

*The production of a reddish-brown flocculent pp.* (ferric hydroxide) *when excess of ammonia is added to a solution is a test for a ferric compound.*

A salt is one of the products obtained by the interaction of acids with (1) metals, (2) metallic oxides, (3) metallic hydroxides, (4) metallic carbonates. The salt obtained is in each case a compound of the metal with the elements of the acid except hydrogen, or, as is usually said, with the *acid radicle.*

---

* It is assumed that the student has become acquainted with the use of chemical formulæ. *s.* ELEMENTARY CHEMISTRY, Chap. VI.

But more than one salt is sometimes obtained by the reaction between a specified acid and the same metallic oxide or hydroxide.

**Exp. 4.**   You are given two exactly equal quantities of a solution of *oxalic acid* in water, and a quantity of a solution of *potassium hydroxide\**.   To one of the quantities of oxalic acid add the potash solution *very cautiously*, from a graduated tube with a stopcock (a *burette*) (Fig. 17) until the liquid is *exactly neutral* (s. *Exp. 2, Chap. VII.*).   Read off, on the burette, the number of cubic centimetres of potash solution added.   To the other quantity of oxalic acid add exactly half as much potash as you have added to the first quantity of the acid.   Evaporate both solutions over *low flames* until a drop taken out on a rod solidifies on cooling ; allow to cool ; collect, dry, and recrystallise the solids from as little water as possible ; collect, dry, and re-crystallise each again.

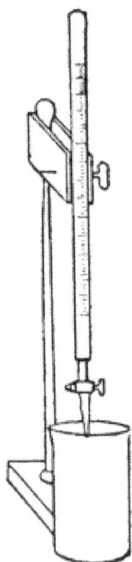

Prove that each solid is (1) a compound of potassium, (2) an oxalate.

*The production of a white pp.* (calcium oxalate) *when calcium chloride is added to a solution, which pp. is insoluble in acetic acid, is a test for the presence of an oxalate in that solution.*

Fig. 17.

Dissolve the solids in distilled water, noting that one is much more soluble than the other, and test each solution with blue and red litmus ; the oxalate of potassium obtained by exactly neutralising oxalic acid with potash is neutral to litmus ; the other oxalate of potassium turns blue litmus red ; this is often expressed by saying that an aqueous solution of the salt has an *acid reaction.*

Quantitative analyses of these oxalates of potassium shew that the composition of the neutral salt is expressed by the formula $K_2C_2O_4$, and the composition of the other salt is expressed by the formula $KHC_2O_4$.

A salt may then be formed from an acid by replacing a part, or the whole, of the hydrogen of the acid by a metal.

---

\* *Note to Demonstrator.*  About 10 grams oxalic acid in 250 c.c. water ; 100 c.c. is enough for each experiment : about 20 grams caustic potash in 250 c.c. water ; 50 c.c. for each experiment.

**Exp. 5.** You are given two exactly equal quantities of a solution of *potash*, in basins, and a quantity of a solution of *sulphuric acid\**. You are also told the weight of each basin.

Fill a burette with the acid; run the acid into one of the quantities of potash until the liquid is exactly neutral; to the other quantity of potash add exactly twice as much sulphuric acid as was required to neutralise the first (equal) quantity of potash. Evaporate each solution to dryness *on a water-bath*; then place each basin on a sand-tray, heap the sand round the basins, and heat them to about 200° for some time. Any free sulphuric acid present will be volatilised at this temperature. Now allow the basins to cool, wipe them carefully, and weigh them. Deduct the weight of each basin from the weight of basin + salt; the difference is the weight of salt obtained.

In one case you obtain a considerably greater weight of salt than in the other. If it is now assumed—and this has been experimentally proved—that sulphuric acid is completely volatilised at a moderate red heat, it follows that you have obtained two distinct *potassium sulphates*. The reasoning is as follows:—

A certain mass, say $x$ grams, of potash (in solution) was exactly neutralised by sulphuric acid; the solution was evaporated; the solid obtained was heated to moderate redness, and weighed; it weighed, say, $x'$ grams.

The same mass ($x$ grams) of potash was neutralised by sulphuric acid, as much sulphuric acid as was used to neutralise the potash was added over and above that required for the neutralisation; the solution was evaporated; the solid obtained was heated to moderate redness, and weighed; it weighed considerably more than $x'$ grams (say $x''$ grams).

(The ratio of $x'$ to $x''$ should be nearly 1 to 1·28.)

But, from the conditions, neither solid obtained could be a mixture of sulphuric acid with potash, or a mixture of potassium sulphate with sulphuric acid; therefore each solid was a sulphate of potassium; therefore there are two potassium sulphates. Confirm this result by dissolving each salt in water and adding a little blue litmus; in one case the litmus is turned red, in the other it remains unchanged.

* *Note to Demonstrator.* About 20 c.c. of a normal solution of potash, as free as possible from carbonate; and about 100 c.c. of a normal solution of sulphuric acid.

Quantitative analysis shews that the compositions of the two salts are expressed by the formulæ $K_2SO_4$ and $KHSO_4$, respectively.

You again conclude that a salt may be formed from an acid by replacing a part, or the whole, of the hydrogen of the acid by a metal (*comp. Exp.* 4).

**Exp. 6.**  You are given two small basins and a quantity of a solution of *hydrochloric acid.* Counterpoise the basins one against the other. It is advisable to label the basins, that in the right hand pan R, that in the left hand pan L.

To each basin add an equal quantity, say 20 c.c., of an aqueous solution of *potash.* Run the solution of hydrochloric acid given you, from a burette, into one of the basins until the liquid is neutral; to the other basin add twice as much of the same hydrochloric acid solution; evaporate the contents of both basins to dryness *on a water-bath*; continue to heat the solids obtained at $100^{\circ}$ *so long as any trace of acid-smelling fumes is given off*; then allow the basins to cool; dry each carefully; place them on the pans of the balance, basin R on the right-hand pan, and basin L on the left-hand pan, and put the counterpoise used at the beginning of the experiment on the same pan as it was then placed on.

The basins and their contents equilibrate each other.

Hence you conclude that the two equal masses of potash have reacted each with the same mass of hydrochloric acid, and that the excess of hydrochloric acid added in one case— i.e. the hydrochloric acid which did not interact with the potash—has been removed by evaporation. Hence, probably, the same salt has been formed in each case.

Dissolve the two solids in the basins in distilled water, and prove that both solutions are *neutral to litmus.* Prove also that both solutions contain (1) a compound of potassium, (2) a chloride (*s. Exp.* 4, *Chap. VII.*; and *Exp.* 8, *Chap. VI.*).

The proof that both solids are the same salt could be rendered complete only by careful quantitative analyses. Such analyses have been made, and have proved that only one salt is obtained by the interaction of an aqueous solution of hydrochloric acid with potash, and that this salt has the composition $KCl$.

The relation between the composition of an acid and that of a salt is, that the salt is a compound of a metal with the elements of the acid except the whole, or a part, of the hydrogen of the acid.

The term *acid radicle* is generally used to denote the elements of an acid minus the replaceable hydrogen of the acid.

You have now gained a fairly clear notion of the essential property of an acid, viz. that it is a compound of hydrogen which reacts with metals, metallic oxides, hydroxides, and carbonates, to form salts. You also know the relation between the composition of a salt and that of the acid from which the salt is obtained. You have learned that some acids react with potash to form more than one salt. You have experimentally examined the meaning of the phrase, *replacing the whole or a part of the hydrogen of an acid by a metal.* Incidentally you have become acquainted with tests by which the presence in solutions of (1) a compound of magnesium, (2) an oxalate, (3) a ferric compound, may be proved. You have also learned how to use a burette.

Acids which react with potash or soda
to form only *one* stable salt  are called *monobasic* acids.

| | | | | | |
|---|---|---|---|---|---|
| ,, | ,, | *two* stable salts | ,, | *dibasic* | ,, |
| ,, | ,, | *three* stable salts | ,, | *tribasic* | ,, |
| ,, | ,, | *n* stable salts | ,, | *n basic* | ,, |

**Exp. 7.**  You are given *carbolic acid*; examine the reactions which occur between this compound and (1) sodium, (2) sodium oxide or hydroxide, (3) sodium carbonate ; and determine from the results of your experiments whether the compound is or is not properly called an acid.

*Reference to* " ELEMENTARY CHEMISTRY." Chap. IX.

# CHAPTER IX.

## CLASSIFICATION OF SALTS.

EXPERIMENTS conducted in Chap. VIII. shewed that hydrochloric acid reacts with potash to produce one salt, and that sulphuric acid and oxalic acid each reacts with potash to produce two salts. Aqueous solutions of one of the potassium sulphates, and one of the potassium oxalates, you prepared, shewed an *acid reaction towards litmus*, i.e. the solutions turned blue litmus red.

The compositions of these salts of potassium, and their actions on litmus, are presented in the following table :—

Potassium chloride.          Potassium sulphates.

KCl. *Neutral to litmus.*     $K_2SO_4$. *Neutral to litmus.*

$KHSO_4$. *Acid to litmus.*

Potassium oxalates.

$K_2C_2O_4$. *Neutral to litmus.*

$KHC_2O_4$. *Acid to litmus.*

There is an evident connexion between the compositions of these salts and their neutrality or non-neutrality towards litmus.

But all salts composed of a metal, an acid radicle, and hydrogen, do not shew an acid reaction towards litmus.

**Exp. 1.** Add *potassium carbonate* to water until no more of the salt dissolves; now pass *carbon dioxide* into the liquid for some time. Crystals separate from the solution : collect these on a filter; wash them by pouring two or three successive *very small quantities of cold water* over them on the filter. Then dissolve the crystals in water and examine the action of the solution on litmus; it is slightly alkaline, i.e. it turns red litmus bluish. The compositions of the two

carbonates of potassium are :—

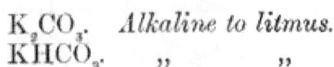

$K_2CO_3$.    *Alkaline to litmus.*
$KHCO_3$.      „         „

Purify *copper sulphate* by recrystallisation from water ; then dissolve the crystals in water, and place a piece of blue litmus paper in the solution ; the liquid has an acid reaction, although the composition of the salt in the solution is $CuSO_4$.

Some metals, and the oxides &c. of these metals, react with a specified acid to form two salts, both of which are compounds of the metal with the elements of the acid except the whole of the hydrogen of the acid. Iron for instance forms two sulphates, which have the compositions $FeSO_4$ and $Fe_2(SO_4)_3$, respectively.

You cannot prove the compositions of these salts except by quantitative analyses; but you may prepare the salts, prove that each is a sulphate of iron, and that the difference of composition is a difference in the ratio of the mass of metal to that of the acid radicle. This may be done as follows.

**Exp. 2.** To a considerable quantity of warm dilute *sulphuric acid**\* in a basin add *iron filings*, little by little, as long as the iron is dissolved; then add a little more iron, evaporate over a low flame to near the *crystallising point* of the solution, filter from undissolved iron into a basin, and allow to cool.

*The crystallising point of a solution is that degree of concentration, for any given temperature, at which the liquid is saturated with the solid held in solution, so that a slight lowering of temperature is attended with deposition of some of the solid.*

Green crystals are deposited as the filtrate cools; collect these, press them repeatedly between porous paper, powder them and press again. Then prove that this solid is (1) a sulphate, (2) a compound of iron.

*To prove that the solid is a compound of iron, make a bead of borax on the end of a clean platinum wire, take a very little of the salt on to this bead and heat in the reducing flame of a Bunsen-lamp until all melts; on cooling a greenish-yellow bead* (borate of iron and sodium) *is obtained.*

Dissolve the remainder of the green crystals of the *sulphate*

* *Note to Demonstrator.* About 200 c.c. of a solution of 1 part concentrated acid in about 10 parts water should be given.

*of iron* you have prepared in dilute *sulphuric acid*, and heat; add a little concentrated *nitric acid* to the hot liquid, a few drops at a time, till the colour is reddish-yellow; now evaporate to dryness over a *low flame.* Collect the yellowish-white solid that remains; free it from adhering liquid by repeated pressure between porous paper; prove (with a small quantity) that it is (1) a compound of iron (*use the borax-bead test*), (2) a sulphate.

[The action of nitric acid was to supply oxygen; when concentrated nitric acid is heated it decomposes, giving oxygen, water, and oxides of nitrogen ($2HNO_3 = H_2O + 2NO_2 + O$).]

Dissolve the remainder of this *sulphate of iron* in dilute *sulphuric acid*; warm the liquid; add some *iron filings*; warm for some time; if all the iron dissolves add more iron, and continue to warm until the liquid is near its crystallising point; then filter; as the filtrate cools it deposits crystals of the green sulphate of iron.*

There are therefore two sulphates of iron; one is composed of more acid radicle relatively to metal than the other.

**Exp. 3.** You are asked to determine whether, when *zinc* and *sulphuric acid* react, more than one zinc sulphate is produced. The details given in *Exp.* 2 will be a sufficient guide to indicate how you ought to proceed.

Note that the only conclusive proof of the production or non-production of more than one sulphate would be quantitative analyses of the solids produced by the interaction of zinc with sulphuric acid under varying conditions.

You can prove by qualitative, and very roughly quantitative, experiments that whether much or little acid is used, the product is a white crystalline zinc sulphate the aqueous solution of which is neutral to litmus. Hence it is probable that only one sulphate of zinc is produced by the interaction of zinc and sulphuric acid.

**Exp. 4.** You are given solutions of (1) *stannous chloride,* $SnCl_2$; (2) *stannic chloride,* $SnCl_4$.

To a portion of each solution add *mercuric chloride* solution and warm; a dark grey pp. (finely divided mercury) is produced in the solution of stannous, but no pp. in the solution of stannic, chloride.

Into a portion of each solution pass *sulphuretted hydrogen* gas and warm; a brown pp. (stannous sulphide SnS) is formed

* See Chapter XVII. on *Reduction.*

in one solution, and a yellow pp. (stannic sulphide $SnS_2$) in the other solution.

Into a portion of the solution of stannous chloride pass *chlorine* gas until the liquid is yellow ; warm for some time, and again pass in chlorine ; warm the liquid until the smell of chlorine is removed. Now prove that the solution gives (*a*) no pp. with mercuric chloride, (*b*) a yellow pp. with sulphuretted hydrogen.

You have converted stannous chloride into stannic chloride, by combining the former with chlorine.

Boil a portion of the solution of *stannic chloride* with *copper* turnings for 5 or 10 minutes ; divide the liquid into three parts ; to one part add mercuric chloride ; into another part pass sulphuretted hydrogen gas ; from the results obtained you conclude that stannous chloride was present in the liquid. To the third part add *ammonia solution* until the liquid smells strongly of ammonia ; a white gelatinous pp. is formed, and the liquid is coloured azure blue :—the pp. is stannous hydroxide ($Sn_2O_3H_2$), *the blue colour shews the presence of a compound of copper.*

Hence you conclude that *stannic chloride* has reacted with copper to form stannous chloride, and that some of the copper at the same time has dissolved. As you know that the compositions of stannous and stannic chlorides are $SnCl_2$ and $SnCl_4$, respectively, you conclude that, probably, the chlorine removed from the stannic chloride by the reaction with copper has combined with copper, and that the copper chloride thus formed has dissolved in the water present. Accurate quantitative investigation confirms this supposition, and shews that the reaction which occurs between a solution of stannic chloride and copper may be thus represented in an equation ;—

$$SnCl_4Aq + Cu = CuCl_2Aq + SnCl_2Aq.$$

The relations between the compositions of these two chlorides of tin and the acid from which both are derived, are similar to the relations between the compositions of the two sulphates of iron and the acid from which they are both derived. Each is a compound of a metal with an acid radicle ; in the case of the iron salts this acid radicle is itself composed of two elements (sulphur and oxygen), in the case of the tin salts the acid radicle is a single element (chlorine). The salt whose name ends in *ic* is a compound of more of the acid radicle, relatively to a fixed mass of the metal, than the salt whose name ends in *ous*. The *-ous* salt was changed to the *-ic* salt by combining with it more of the acid radicle ; the *-ic* salt was changed to the *-ous* salt, in one case by combining with it more of the metal, in the other case by removing part of the acid radicle.

A salt which is formed from acids by replacing the whole of the replaceable hydrogen by a metal is usually called a *normal salt*; a salt which is formed from an acid by replacing a portion of the replaceable hydrogen by a metal is usually called an *acid salt*. The salts $FeSO_4$, $Fe_2(SO_4)_3$, $SnCl_2$, $SnCl_4$, $K_2SO_4$, $K_2C_2O_4$, $K_2CO_3$, are normal salts. The salts $KHC_2O_4$, $KHSO_4$, $KHCO_3$, are acid salts. It is to be noted that most acid salts shew an acid reaction towards litmus, but that some acid salts (as defined above) do not exhibit an acid reaction some towards litmus. Most normal salts are neutral to litmus ; however are acid and some are alkaline. (*s. Exp.* 1 *of this Chap.*)

Many normal salts may be formed by the interaction of an acidic with a basic oxide.

**Exp. 5.** Pass a stream of dry *carbon dioxide* over some warm powdered *lime* (calcium oxide) in a glass tube (Fig. 18). After a time stop the current of the gas, remove the white

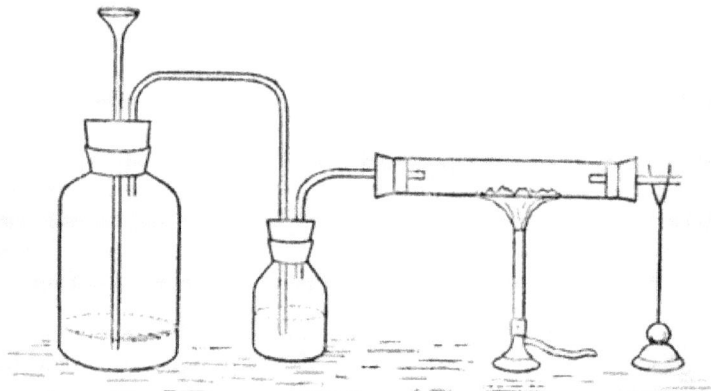

Fig. 18.

solid from the tube and prove that it is (1) a carbonate, (2) a compound of calcium.

*Calcium compounds when volatilised in a non-luminous flame give a red colour to the flame. Solutions of calcium compounds neutralised with ammonia, give a white pp.* (calcium oxalate) *on addition of a solution of ammonium oxalate, which pp. is insoluble in acetic acid.*

The chemical change which you have conducted is represented in an equation thus, $CaO + CO_2 = CaCO_3$. Calcium carbonate is a salt, calcium oxide is a basic oxide, and carbon dioxide is an acidic oxide.

**Exp. 6.** Melt some solid *potassium oxide* in a silver (or nickel) basin; add a *little* powdered *manganese dioxide*, and continue to heat for some minutes; then allow to cool and dissolve the greenish solid in the basin in cold water. A deep green solution is obtained. This solution contains the salt *potassium manganate* ($K_2MnO_4$). The reaction which occurred between the potassium oxide—a basic oxide—and the manganese dioxide—an acidic oxide—is thus represented in an equation;—

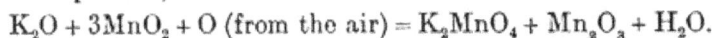

$$K_2O + 3MnO_2 + O \text{ (from the air)} = K_2MnO_4 + Mn_2O_3 + H_2O.$$

**Exp. 7.** Potassium oxide is a basic oxide; chromium trioxide is an acidic oxide. Heat together, in a crucible, *potassium oxide* and *chromium trioxide* ($CrO_3$); when the whole mass has been molten for a few minutes allow to cool, and dissolve in water; evaporate the yellow solution to the crystallising point; cool; collect and purify the yellow crystals which form. Set these crystals aside; call them $A$.

To a solution in water of chromium trioxide add a solution in water of potassium oxide till the liquid is as nearly as possible neutral; evaporate, collect and purify the yellow crystals which form; call these crystals $B$.

Prove that $A$ and $B$ are both (1) compounds of potassium, (2) chromates.

*The production of a red pp. of silver chromate* ($Ag_2CrO_4$), *soluble in hot concentrated nitric acid, when a solution of silver nitrate is added to an aqueous solution of a salt, is a test for a chromate.*

As far as your experiments indicate, $A$ and $B$ are the same compound. Quantitative analysis establishes this conclusion. The composition of $A$ and $B$ is represented by the formula $K_2CrO_4$: the reactions by which you have prepared this salt may be thus represented:—

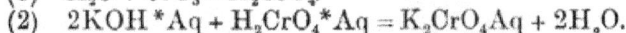

(1)   $K_2O + CrO_3 = K_2CrO_4$.
(2)   $2KOH*Aq + H_2CrO_4*Aq = K_2CrO_4Aq + 2H_2O$.

* An aqueous solution of $K_2O$ contains the alkali KOH ; an aqueous solution of $CrO_3$ contains the acid $H_2CrO_4$.

Many normal salts may then be regarded as, (1) metallic derivatives of acids, (2) compounds of basic with acidic oxides.

Some salts are formed by the combination of basic oxides, or hydroxides, with normal salts : such salts are usually called *basic salts*.

**Exp. 8.**   The compound bismuth chloride ($BiCl_3$) is a *normal salt* ; bismuth oxide ($Bi_2O_3$) is a basic oxide.  Melt some *bismuth chloride* in a basin over a Bunsen-lamp, and add little by little some powdered *bismuth oxide*.  The white solid thus formed is *bismuth oxychloride* ; its composition is expressed as that of a compound of the normal salt bismuth chloride with the basic bismuth oxide ;—$BiCl_3 . Bi_2O_3$.  This compound is a *basic salt*.

**Exp. 9.**   Lead acetate $[Pb(C_2H_3O_2)_2]$ is a *normal salt* ; lead oxide (PbO) is a basic oxide.  Dissolve about 10 grams *lead acetate* in about 50 c.c. water ; boil ; add about 7 grams *lead oxide* ; keep boiling till the greater part of the lead oxide has dissolved ; then filter, and allow filtrate to cool.  The *basic salt* $Pb(C_2H_3O_2)_2 . PbO$ is deposited as the liquid cools.

In this chapter you have learned more concerning the relations between the composition of salts, and the composition, on the one hand, of the acids from which salts are derived, and on the other hand, of the oxides by the combination of which salts are frequently produced.

You have gained some knowledge of the meaning of the terms used in the classification of salts, *normal, acid, basic salts*.  You have found that some metals form two series of salts by interacting with one and the same acid ; and that, in these cases, the ratio of acid radicle to metal is different in the two series of salts.  You have also learned how to pass from one of these series to the other, in the cases of some iron and tin salts.

Incidentally you have performed tests by which the presence in liquids of compounds of (1) iron, (2) copper, (3) calcium, and (4) the presence of a chromate, may be proved.

*Reference to* "ELEMENTARY CHEMISTRY." Chaps. IX. and XI.

# CHAPTER X.

**Exp. 1.** To two quantities of distilled water (about 100 c.c. each) in basins add *sodium* and *potassium*, respectively. Cut the sodium and potassium into *small pieces*, and add these one at a time, until about 5 grams of each metal has been added.

*Never touch sodium or potassium with wet hands.*

Test a little of each solution with litmus; both solutions turn red litmus blue. These solutions contain the compounds NaOH and KOH, respectively. [$M + H_2O + Aq = MOHAq + H$; $M = Na$ or $K$.]

**Exp. 2.** Into each of two test tubes pour a little solution of *ferric chloride*; into two other tubes, a solution of *manganese sulphate*; into two others, a solution of *zinc sulphate*; and into two others, a solution of *mercuric chloride*. To one of each pair of solutions add some of the solution of *potassium hydroxide*, and to the other some of the *sodium hydroxide* solution, you prepared in Exp. 1.

In every case a pp. is produced; observe that addition of a considerable quantity of the sodium or potassium hydroxide to the liquid containing zinc sulphate brings about a solution of the pp. which was at first produced.

The pps. produced have the following compositions;—

$$Fe_2O_6H_6, \quad MnO_2H_2, \quad ZnO_2H_2, \quad HgO.$$

The chemical changes which occurred may be represented thus;—

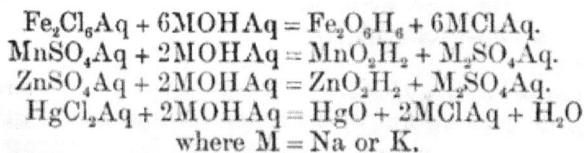

$$Fe_2Cl_6Aq + 6MOHAq = Fe_2O_6H_6 + 6MClAq.$$
$$MnSO_4Aq + 2MOHAq = MnO_2H_2 + M_2SO_4Aq.$$
$$ZnSO_4Aq + 2MOHAq = ZnO_2H_2 + M_2SO_4Aq.$$
$$HgCl_2Aq + 2MOHAq = HgO + 2MClAq + H_2O$$
$$\text{where } M = Na \text{ or } K.$$

In each case reaction occurred between a salt of a heavy metal and a solution of an alkali ; in three cases an hydroxide of the heavy metal was pptd., in the other case an oxide of the heavy metal was pptd. ; in every case a salt of the metal of the alkali was formed.

**Exp. 3.** Into portions of the solutions of *potassium and sodium hydroxides* prepared in Exp. 1, pass *carbon dioxide* for a few minutes. The carbon dioxide reacts with the alkali present to form carbonate of sodium or potassium. To prove that a carbonate is formed, pour each solution into a small flask with a well-fitting cork carrying tubes as shewn in fig. 19. Let the exit tube from each flask pass into a clean beaker, in which you place a little lime water. Pour a little hydrochloric or sulphuric acid through the funnel tube into each flask. A gas comes off which reacts with the lime water to produce a white pp.; this gas therefore is carbon dioxide. But carbonates are recognised by the fact that they interact with acids to evolve carbon dioxide gas.

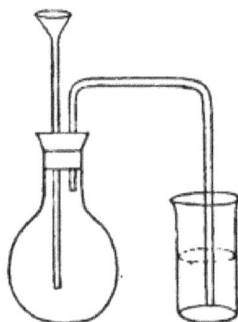

Fig. 19.

Carbon dioxide is an acidic oxide. Alkalis in solution react with many acidic oxides to produce salts.

**Exp. 4.** Place two small pieces of *suet*, each weighing about 50 grams, in basins ; into one basin pour a solution of about 10 grams of *potassium hydroxide*, and into the other a solution of about 10 grams of *sodium hydroxide*. Heat the contents of the basins with constant stirring until clear liquids are obtained ; continue heating for 10—15 minutes ; then allow to cool. A semi-solid crust forms on the surface of each liquid. Break the crusts and pour out the liquids. The crusts are *soaps* ; in the one case potash-soap, in the other case soda-soap. The liquids contain *glycerin*; prove this by moistening a borax bead with the liquid and bringing the bead into the upper part of a Bunsen-flame, when a green colour is imparted to the flame.

Alkalis react with fats to form soaps and glycerin. The change of composition which occurs is similar to that noticed in Exp. 2. Both classes of change may be represented by the general equation ; $RX + MOH = ROH + MX$, where RX is a compound of a positive and a negative radicle (a salt), and MX is a compound of the metal of the alkali used with the negative radicle of the salt RX. In Exp. 2 the compounds RX were metallic salts; R was Fe, Mn, or Zn, X was $SO_4$ or Cl. In the present Exp. RX is a fat ; both R and X are composed of more than one element. Common mutton suet is almost pure *stearin* or *glyceryl stearate*. The reaction which occurs when this fat is *saponified* by potash may be expressed thus ;—

$$(C_3H_5)(C_{18}H_{35}O_2)_3 + 3KOH = C_3H_5.O_3H_3 + 3K.C_{18}H_{35}O_2$$

   glyceryl stearate          glycerin   potassium stearate
                                                (potash-soap)

[Here $R = C_3H_5$, and $X = C_{18}H_{35}O_2$.]

Recall *Exp. 2, Chap. VII.* where you neutralised acids by the interactions of these acids with solutions of potassium and sodium oxides. Solutions of these oxides contain the alkalis potassium and sodium hydroxide [$M_2O + H_2O + Aq = 2MOHAq.$].

The hydroxides of the metals lithium, sodium, potassium, rubidium, and cæsium, are alkalis.

Alkalis then have the composition MOH where $M = Li$, Na, K, Rb, or Cs; they are very soluble in water; aqueous solutions of alkalis neutralise acids forming salts and water ; they react with many acidic oxides to form salts and water ; they react with solutions of salts of many heavy metals to form pps. of hydroxides, or sometimes oxides, of these metals, and at the same time to produce compounds of the metal of the alkali used with the acid radicle of the salt of the heavy metal ; they saponify fats.

The *hydroxides of the metals calcium, strontium, and barium*, resemble the alkalis in most of their properties : the composition of these hydroxides is $MO_2H_2$ where $M = Ca$, Sr, or Ba.

**Exp. 5.**   Prove experimentally (1) that aqueous solutions of *calcium* and *barium hydroxides* are alkaline to litmus ; (2) that these solutions react with solutions of *ferric chloride, manganese sulphate, zinc sulphate*, and *mercuric chloride* in a way similar to that in which the alkalis react ; (3) that these solutions react with *carbon dioxide* to form carbonates of calcium and barium respectively ; (4) that these solutions very slightly and slowly *saponify fat* ; (5) that the solutions *neutralise acids* forming salts.

4—2

**Exp. 6.**  Weigh out approximately equal masses of the alkali *caustic potash* (KOH), and the alkaline hydroxide *caustic baryta* (BaO$_2$H$_2$), say about a couple of grams of each, and add each to the same quantity of *water*, say to 10 c.c.  The alkali dissolves rapidly and a great deal of heat is produced during the solution ; only a little of the alkaline hydroxide dissolves, and comparatively very little heat is produced during the solution.

**Exp. 7.**  Heat a little solid *dry caustic potash*, and a little solid *dry calcium hydroxide*, in separate dishes, covering each with a dry funnel ; the potash melts at a high temperature but does not give off water ; the calcium hydroxide does not melt, but it is decomposed into calcium oxide and water.

Now tabulate the results of your experiments on alkalis and alkaline hydroxides, shewing clearly the resemblances and differences between these compounds.

Recall *Exp.* 8, *Chap. VIII.*, in which you proved that ferric hydroxide reacts with acids to form salts ; ferric hydroxide is insoluble in water ; it is neutral to litmus ; it does not saponify fats ; it does not react with carbon dioxide to form a carbonate of iron.

The term *basic hydroxide* may be used to include all hydroxides which interact with acids to form salts.

You have now learned to contrast acidic with basic oxides ; and to contrast alkalis, alkaline hydroxides, and neutral hydroxides, with acids.  You have learned some of the relations of composition and properties between salts, acids, basic hydroxides (including alkalis, alkaline hydroxides, and neutral hydroxides), basic oxides, and acidic oxides.

The term *base* is frequently used to signify a compound which interacts with acids to form salts and water, only, the term includes basic oxides and basic hydroxides, and also some other compounds, such as ammonia (NH$_3$), ethylamine (NH$_2$.C$_2$H$_5$), &c.

*Reference to* "ELEMENTARY CHEMISTRY"; Chaps. VIII. and IX.

# CHAPTER XI.

## REACTIONS BETWEEN ACIDS AND SALTS.

**Exp. 1.** To the salt *barium chloride* add a little concentrated *sulphuric acid*. Arrange the experiment as shewn in fig. 20. Put about 3 or 4 grams of the salt in the small flask, and pour the acid down the funnel tube. Allow the gas which comes off to pass into a little distilled water; the gas dissolves rapidly. When the reaction has moderated, heat the flask gently and continue to heat for 5 minutes or so.

*Remove the exit tube from the water before you take away the lamp.*

When the contents of the flask are cold add some water; shake up well; pour off the water; add more water, and shake up again; pour

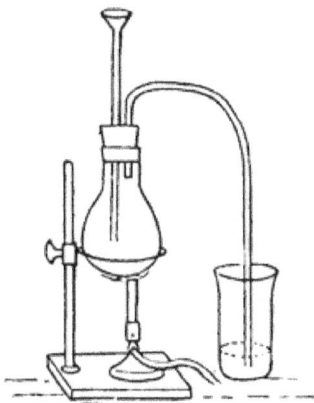

Fig. 20.

off; continue to wash the white solid by decantation until the washings are no longer acid to litmus paper. Now prove that the white solid is *barium sulphate* (s. *Chap. VII. Exp.* 5). Prove that the aqueous solution of the gas produced in the reaction, (1) is acid to litmus, (2) gives the reaction with silver nitrate characteristic of a chloride, (3) neutralises caustic potash forming potassium chloride, (4) dissolves metals—say zinc and magnesium—with evolution of hydrogen. The liquid is an aqueous solution of *hydrochloric acid*.

The reaction which you have carried out is represented in an equation thus ;—

$$BaCl_2 + H_2SO_4 = BaSO_4 + 2HCl.$$
$$\text{salt} \qquad \text{acid} \quad \text{new salt} \quad \text{new acid.}$$

By the interaction of a salt with an acid there have been produced, (1) a new salt, composed of the metal of the original salt and the radicle of the acid used ; (2) a new acid, composed of the acid radicle of the original salt and the hydrogen of the acid used.

**Exp. 2.** In a flask, arranged with a cork and tubes as in Exp. 1, place 3—5 grams of the salt *sodium bromide*, and add a quantity of a very concentrated solution of *phosphoric acid*. Warm the flask, and lead the colourless acid-smelling gas which comes off into distilled water. Proceed as directed in Exp. 1. Prove that the water into which the gas was conducted contains an acid, by shewing that it turns blue litmus red, dissolves metals with evolution of hydrogen, and neutralises caustic potash ; prove that the acid is *hydrobromic acid, by adding silver nitrate and obtaining a very pale yellow pp. of silver bromide* ($AgBr$), *soluble with difficulty in a considerable quantity of ammonia solution.* Prove that the solid obtained by evaporating the liquid in the flask to dryness and heating till acid fumes cease to be evolved is, (1) a compound of sodium, (2) a phosphate.

The reaction which occurred in the flask is represented in the following equation :—

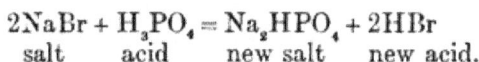

$$2NaBr + H_3PO_4 = Na_2HPO_4 + 2HBr$$
$$\text{salt} \qquad \text{acid} \qquad \text{new salt} \qquad \text{new acid.}$$

This reaction is analogous to that which occurred between sodium chloride and sulphuric acid.

**Exp. 3.** Conduct an experiment similar to 1 and 2, but put the salt *potassium nitrate* in the flask and add concentrated *sulphuric acid*.

Lead the gas evolved into water ; prove that the solution contains nitric acid.

Evaporate the solution in the flask to dryness *in the draught cupboard*, heat the residue as long as acid fumes come off, and prove that the residual solid is potassium sulphate.

Nitric acid is colourless ; you noticed a reddish-brown gas

in the flask while the salt and the acid were heated together. What is this gas?

**Exp. 4.** Heat in a test tube a little *concentrated nitric acid* and a little *concentrated sulphuric acid*; a reddish-brown gas is evolved. Put a cork with an exit tube into the test tube and lead a little of the gas into a solution of ferrous sulphate acidified with sulphuric acid; a brown-black liquid is formed (comp. test for nitric acid, *Exp.* 4, *Chap. VII.*).

Now lead some of the gas produced in Exp. 3 into a cold acidulated solution of ferrous sulphate.

From the result obtained you conclude that, probably, the gas obtained in Exp. 4 is the same as the reddish-brown gas produced along with nitric acid in Exp. 3. The gas in question is a mixture of nitric oxide (NO) and nitrogen dioxide ($NO_2$).

The reaction which occurred in Exp. 3 may be thus represented in equations ;—

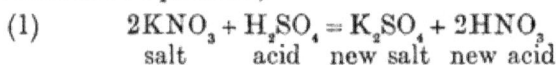

(1) $$2KNO_3 + H_2SO_4 = K_2SO_4 + 2HNO_3$$
salt     acid     new salt   new acid

(2) $2HNO_3 + H_2SO_4 = H_2O.H_2SO_4 + 2NO + 3O$ (some of the NO and O combine to form $NO_2$).

A portion of the new acid produced in the primary change reacts with some of the acid used to decompose the salt, and is thereby itself decomposed.

**Exp. 5.** Allow the salt *sodium carbonate* to react with *nitric acid*; prove that carbon dioxide is evolved and that sodium nitrate remains in the flask (evaporate the contents of the flask to dryness on a water-bath and heat at 100° until acid fumes cease to come off). The reaction is thus represented :

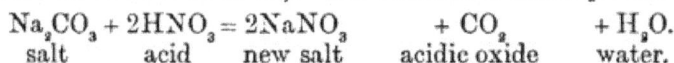

$$Na_2CO_3 + 2HNO_3 = 2NaNO_3 \qquad + CO_2 \qquad + H_2O.$$
salt     acid     new salt     acidic oxide    water.

A new acid is not formed, but an acidic oxide and water are produced. Reasoning from the results of Exp. 1 and 2 we should have expected carbonic acid ($H_2CO_3$) to be produced in the present exp. ; this acid, if it exists, is extremely unstable and is very easily decomposed into an oxide and water.

In Exp. 3 an acid was produced, part of which was decomposed under the experimental conditions ; in the present Exp. an acid has probably been produced, but at the moment of its production it has been decomposed into an oxide and water.

**Exp. 6.**   Allow moderately concentrated *sulphuric acid* to react with a little *potassium iodide*; lead the gas produced into water; prove (1) that the solution contains an acid; (2) that this acid is *hydriodic, by adding silver nitrate and obtaining a pale yellow pp. of silver iodide* (AgI) *insoluble in ammonia solution.*   Prove that the residue in the flask, after removing the excess of sulphuric acid by evaporation and heating, is potassium sulphate.

Notice that a greyish violet solid is formed in the neck of the flask.  Remove a little of this solid to a test tube: prove that it is slightly soluble in water; and *very soluble in alcohol, giving a brown liquid, a drop or two of which added to a very little starch, which has been boiled with a good deal of water and allowed to cool, produces a deep blue colour* (so-called iodide of starch).  These reactions prove that the solid is iodine.

In this reaction a new acid (hydriodic) has been produced; but part of this acid interacts with the excess of sulphuric acid present to produce iodine &c.   Thus

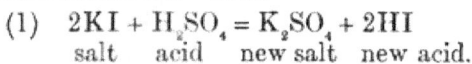

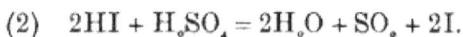

$$(1)\quad 2KI + H_2SO_4 = K_2SO_4 + 2HI$$
$$\text{salt}\quad\text{acid}\quad\text{new salt}\quad\text{new acid.}$$

$$(2)\quad 2HI + H_2SO_4 = 2H_2O + SO_2 + 2I.$$

In this chapter you have learned that the normal products of the reaction between an acid and a salt are, (1) a new salt, composed of the metal of the original salt and the radicle of the acid used; and (2) a new acid, composed of the acid radicle of the original salt and the hydrogen of the acid used.  You have learned that secondary changes frequently occur, generally resulting in a change of part, or even the whole, of the new acid produced into less complex compounds or into its elements.

Incidentally, you have learned tests for an iodide and a bromide.

**Exp. 7.**   You are given the salt *ethylic acetate* $(C_2H_5.C_2H_3O_2)$; determine whether the reaction which occurs when this salt is heated with *hydrobromic acid* follows the course of the normal reaction between an acid and a salt.  You are told that *ethylic bromide* is a liquid about $1\frac{1}{2}$ times heavier than, and insoluble in, water; you are also told that the test for *acetic acid consists in exactly neutralising with ammonia and adding ferric chloride solution, when a red colour (due to formation of ferric acetate) is produced.*

*Reference to* "ELEMENTARY CHEMISTRY"; Chaps. IX. and XI.

# CHAPTER XII.

The elements chlorine, bromine, and iodine are placed in the same class.

**Exp. 1.** Mix about 2 grams of *potassium chloride* with about 3 grams of *manganese dioxide;* place the mixture in a flask fitted with a cork, exit tube and funnel tube, as shewn in fig. 20, (p. 53); let the exit tube pass into a dry jar; set the apparatus in the draught place ; pour a little concentrated *sulphuric acid* diluted with its own weight of water down the funnel tube into the flask, and warm gently.

A greenish-yellow gas collects in the jar. This gas is *chlorine*.

**Exp. 2.** Conduct an experiment similar to Exp. 1, but use *potassium bromide* in place of potassium chloride.

A reddish-brown gas, condensing to dark-red drops, collects in the jar. This liquid is *bromine*.

**Exp. 3.** Place a mixture of about 2 grams *potassium iodide* and 3 grams *manganese dioxide* in a small retort

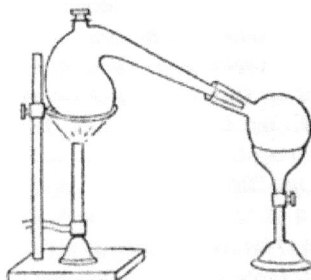

Fig. 21.

arranged so that the beak passes into a small dry flask (Fig. 21). Add some *sulphuric acid* and warm.

A violet coloured vapour appears, and condenses on the colder parts of the retort, and in the flask, to a lustrous, greyish-violet, solid.  This solid is *iodine*.

Chlorine, bromine, and iodine, are prepared from similar compounds under similar conditions.  The reactions which occur in Exps. 1, 2, and 3 may be thus represented ;

$$2KX + MnO_2 + 2H_2SO_4 = K_2SO_4 + MnSO_4 + 2H_2O + 2X$$
where $\qquad\qquad\qquad X = Cl, Br,$ or I.

**Exp. 4.**  Place about 20 grams *potassium chloride* in a flask arranged as shewn in fig. 20 (p. 53).  Pour concentrated *sulphuric acid* into the flask ; collect (by downward displacement) the colourless, strongly acid-smelling, gas which is produced in *dry* jars ; fill three jars with the gas ; cover each jar with a *well greased* glass plate and set it aside.

The gas produced is *hydrogen chloride* (HCl).  The reaction which proceeds in this Exp. is

$$KCl + H_2SO_4 = KHSO_4 + HCl.$$

**Exp. 5.**  Place a crystal or two of *potassium iodide* in a test tube, add a little concentrated *sulphuric acid* and warm ; violet vapours condensing to a greyish-violet solid, are produced. *Iodine* is evidently produced.  But chlorine was not produced in the similar experiment with potassium chloride and sulphuric acid.

**Exp. 6.**  Repeat Exp. 5, but use a crystal of *potassium bromide* in place of potassium iodide.  *Bromine* is formed.

**Exp. 7.**  You are given concentrated aqueous solutions of *hydriodic acid* (HI) and *hydrobromic acid* (HBr); to a little of each add a little concentrated *sulphuric acid*, and warm ; in one case *iodine*, and in the other case *bromine*, is formed.

The results of Exps. 1 to 3, and of Exp. 4 would lead you to suppose that hydrogen iodide would have been produced in Exp. 5, and hydrogen bromide in Exp. 6 ; but the results of Exp. 7 shewed that if these compounds were produced by the interaction of warm concentrated sulphuric acid with potassium iodide and bromide, respectively, they would be decomposed.

**Exp. 8.**  To small quantities of the solutions of *hydriodic* and *hydrobromic acids* used in Exp. 7 add some syrupy solution of *phosphoric acid* (solution of phosphoric acid—$H_3PO_4$—

so concentrated as to be viscid), and warm : neither iodine nor
bromine is formed.    Therefore it may be possible to procure
hydrogen iodide and bromide by reactions between potassium
iodide, or bromide, and phosphoric acid.

**Exp. 9.**    Heat together small quantities of a very concentrated solution of *phosphoric acid* and (1) *potassium
bromide* (2) *potassium iodide.*    In each case conduct the
experiment in a test tube fitted with a cork and exit tube
passing into a dry test tube.    A colourless acid-smelling gas
is produced in both experiments.    These gases are *hydrogen
bromide* (HBr), and *hydrogen iodide* (HI), respectively.    The
reactions which proceed in this Exp. are

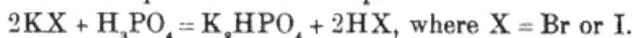

$$2KX + H_3PO_4 = K_2HPO_4 + 2HX, \text{ where } X = Br \text{ or } I.$$

These compounds are more easily produced by decomposing
phosphorus bromide or iodide by water ;

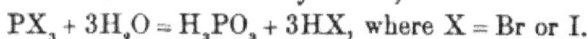

$$PX_3 + 3H_2O = H_3PO_3 + 3HX, \text{ where } X = Br \text{ or } I.$$

**Exp. 10.**    Arrange a *dry* flask as shewn in Fig. 22.
Place about 2 grams of red amorphous *phosphorus* and about

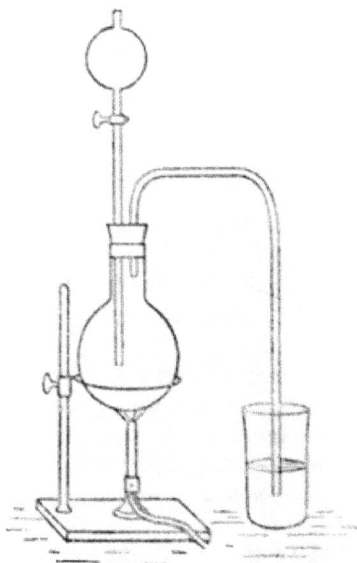

Fig. 22.

15 grams of *iodine* in the flask ; heat gently ; the two elements
combine to **form** nearly black *phosphorous iodide* ($PI_3$).    Place

the cork with its tubes in the flask ; nearly fill the funnel tube with *water*, and allow the water to flow drop by drop into the flask ; the decomposition $PI_3 + 3H_2O = 3HI + H_3PO_3$ proceeds; the *hydrogen iodide* passes off as a gas, the phosphorous acid remains in the flask. Fill three dry jars with the gas ; cover them with *well greased* plates, and set them aside *in the dark.*

**Exp. 11.** Repeat Exp. 10 but use *bromine* in place of iodine. A colourless, fuming, irritant, gas is produced. This gas is *hydrogen bromide.* Fill three jars with the gas, cover them with *well greased* plates, and set aside.

You must now compare the properties of the three gases, hydrogen chloride (HCl), hydrogen bromide (HBr), and hydrogen iodide (HI), which you have prepared.

**Exp. 12.** Bring a bottle of each gas, mouth downwards, under water, and remove the plates ; the water rapidly rises in each jar. After a few minutes slip the plates under the mouths of the jars, and reverse them.

Examine the three liquids with litmus ; each turns blue litmus red.

Neutralise a portion of each solution by *caustic potash* ; evaporate the liquids to dryness at 100°, and heat the residues for a few minutes. Prove that each of the white solids thus obtained is a compound of potassium, and that one is a chloride, one a bromide, and the third an iodide.

The reactions which have occurred are represented thus ;—

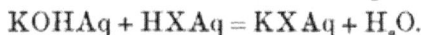

$$KOHAq + HXAq = KXAq + H_2O.$$

The test for an iodide or bromide is based on the facts ; (1) that either is decomposed by chlorine giving a chloride, and iodine or bromine ; (2) that iodine and bromine dissolve very readily in carbon disulphide and colour this liquid violet (iodine) or reddish-brown (bromine).

*To test for an iodide or bromide dissolve in a little water, add a little chlorine-water, shake up, then add a little carbon disulphide and shake again; the carbon disulphide sinks to the bottom of the tube without mixing with the liquid; if it is coloured violet, an iodide was present, if reddish brown, a bromide was present, in the original liquid.* (*Comp. Chap. XI. Exps.* 206.)

**Exp. 13.** Into jars of *hydrogen chloride, hydrogen bromide,* and *hydrogen iodide,* respectively, pour a little concentrated *nitric acid*; notice that the bromide and iodide are decomposed with production of bromine and iodine, respectively, but that the chloride is unchanged.

**Exp. 14.** Expose jars of the three gases to *direct sunlight*; the bromide and iodide are slowly decomposed, with separation of bromine and iodine, the chloride is unchanged.

The three compounds HCl, HBr, HI, are then produced from similar compounds under somewhat similar conditions; the compounds all dissolve in water to form acid solutions which react with alkalis to form similar salts.

The compound HCl is much more stable as regards the action of sunlight and oxidising agents such as nitric acid, than either HBr or HI.

**Exp. 15.** Place a saturated aqueous solution of *hydro-chloric acid* in a V shaped tube; roll two copper wires round pieces of hard gas carbon; immerse these pieces of carbon in

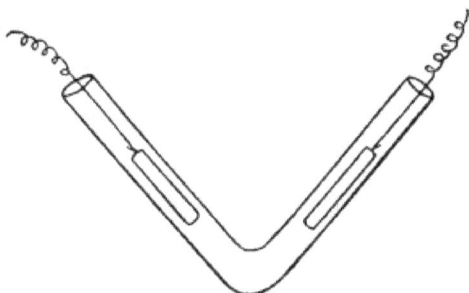

Fig. 23.

the liquid and attach the other ends of the wires to the plates of a battery (Fig. 23). Hydrogen is evolved at the negative electrode, and after some time chlorine at the positive electrode.

**Exp. 16.** Place a very little *mercury* in a bulb-tube of hard glass; connect one end of the tube with a chlorine-generating apparatus, and in the other end place a cork carrying an exit tube passing into a dry flask. *Set the whole apparatus in the draught cupboard.* Pass a slow stream of *chlorine* over the mercury while you warm the latter: a white solid is produced and collects in the exit tube and small flask.

Mercury and chlorine combine directly to form mercuric chloride ($HgCl_2$).

*The fumes of mercury and of mercury compounds are very poisonous.*

**Exp. 17.** Put two very small quantities of *mercury* in two clean mortars; to one add a few crystals of *iodine* and a drop or two of alcohol; to the other add 2 or 3 drops of *bromine*. Rub the contents of each mortar together with a pestle.

The iodine and bromine combine with the mercury to form red *mercuric iodide* ($HgI_2$), and white *mercuric bromide* ($HgBr_2$), respectively.

Chlorine, bromine, and iodine, combine directly with many elements to form compounds having similar compositions, and, generally, similar properties.

**Exp. 18.** You are given solutions of *antimony chloride* ($SbCl_3$) in hydrochloric acid, *antimony bromide* ($SbBr_3$) in hydrobromic acid, and *antimony iodide* ($SbI_3$) in hydriodic acid; pour a little of each solution into a large quantity of water and notice the pps. which are produced. These pps. are *oxychloride*, *oxybromide*, and *oxyiodide, of antimony*, respectively; their compositions, and the conditions under which they are obtained, are similar; thus $SbX_3 + H_2O + Aq = SbOX + 2HXAq$ where $X = Cl$, $Br$, or $I$.

The reactions between aqueous solutions of alkalis and the three elements, chlorine, bromine, and iodine, are similar; compounds of similar compositions and similar properties are produced under similar conditions.

**Exp. 19.** Into a cold solution of the alkali *potassium hydroxide* pass *chlorine* until the liquid smells of chlorine; then add a drop or two of potassium hydroxide solution. The solution now contains *potassium chloride* (KCl) and *hypochlorite* (KClO). The latter salt is decomposed by acids with evolution of chlorine which bleaches a piece of madder-dyed cloth. Place a small piece of madder-dyed cloth in the solution you have just prepared and add a little hydrochloric acid; the cloth is slowly bleached.

**Exp. 20.** To a cold solution of *potassium hydroxide* add *bromine*, drop by drop, until the liquid has a faint yellow colour; then add two or three drops of potassium hydroxide solution ; immerse a piece of madder-dyed cloth in the liquid and add a little hydrochloric acid; the cloth is slowly bleached.

The reactions which occurred in Exps. 19 and 20 may be represented thus ;—

$$2KOHAq + 2X = KXAq + KXOAq + H_2O.$$

**Exp. 21.**   Perform an experiment similar to 19 and 20, but use *iodine* in place of chlorine or bromine : the liquid which is produced does not bleach.   No compound of iodine analogous to KClO and KBrO has been obtained.

**Exp. 22.**   *Perform this experiment in the draught-cupboard.*   Make an aqueous solution of the alkali *potassium hydroxide* (3 parts of the alkali to one part of water); divide it into three parts and heat two of these to boiling. Into one portion pass *chlorine* until the liquid smells of the gas; then evaporate to about half the bulk and set aside to cool.   To another portion of the boiling potash add *iodine*, little by little, until the liquid remains slightly yellow ; evaporate to about half the bulk, and set aside to cool. To the third (cold) portion of the potash solution add *bromine* until the liquid remains distinctly reddish yellow; heat to boiling; add a few drops of bromine ; evaporate to about half the bulk, and set aside to cool.   The three liquids deposit white crystals, mixtures of *potassium chloride* and *chlorate*, *bromide* and *bromate*, *iodide* and *iodate*, respectively. The reactions which have occurred are these ; $6KOHAq + 6X$ $= 5KX + KXO_3 + 3H_2O.$

Potassium chloride is much more soluble in cold water than potassium chlorate ; potassium iodide is soluble, but potassium iodate is nearly insoluble, in alcohol ; potassium bromate is more soluble in alcohol than potassium bromide.   On these facts may be based methods for obtaining approximately pure potassium chlorate, bromate, and iodate, from the mixtures of these salts with chloride &c. prepared in the earlier part of this Exp.

Pour off the mother liquor from each crop of crystals.

Wash the crystals obtained by using chlorine repeatedly with *small* quantities of cold water, continuing this process until the washings cease to give the reaction with silver nitrate characteristic of a chloride.   Dry the crystals which remain, label them $KClO_3$, and set aside.

*Very slightly* warm the two other crops of cystals with

moderately strong alcohol. Repeat the treatment in the case of
the crystals obtained by using iodine until a fresh quantity
of alcohol dissolves only a very little of the solid ; then dry
the residue, label it $KIO_3$, and set aside.

Repeat the treatment with cold alcohol two or three times
in the case of the crystals obtained by using bromine; then
evaporate the alcoholic liquid on a water bath to about half
its bulk ; allow to cool ; collect and dry the crystals which
form, label them $KBrO_3$, and set aside.

Examine the appearance of the three salts you have prepared;
prove that each is soluble in a large quantity of cold water ;
that each is decomposed by heating in a dry tube, with evolution of oxygen, and production of a white residue which is
potassium chloride, bromide, or iodide, according as it is
obtained from potassium chlorate, bromate, or iodate.

The decomposition of the chlorate &c. by heat, is represented thus ; $KXO_3 = KX + 3O$, where $X = Cl, Br, I$.

The three elements chlorine, bromine, and iodine evidently
exhibit similar chemical properties ; they form similar compounds under similar conditions.

*Reference to* " ELEMENTARY CHEMISTRY " Chap. XI, pars.
149—159.

# CHAPTER XIII.

THE elements magnesium, zinc, and cadmium are placed in the same class.

**Exp. 1.** Observe the prominent physical properties of the three elements; their *hardness, malleability, lustre,* and *colour.* Prove that each is a good *conductor of electricity* by placing each in a circuit including an electric bell, as shewn in fig. 24.

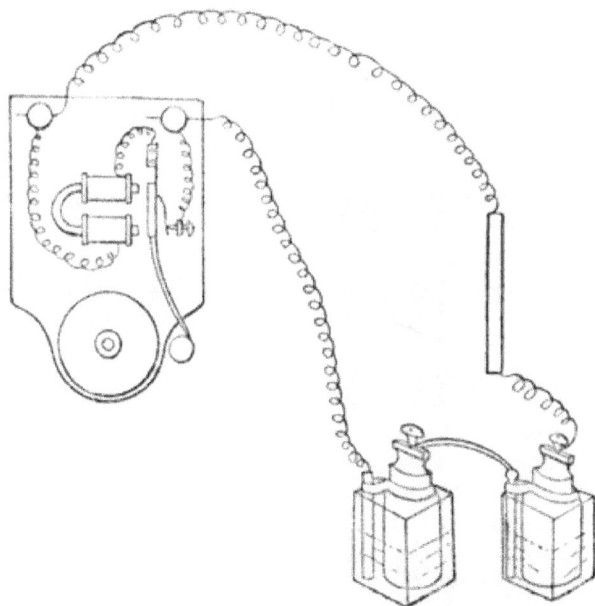

Fig. 24.

The ringing of the bell proves that the current is passing through the magnesium, **zinc,** or cadmium.

**Exp. 2.** Pass an electric current through aqueous solutions of (1) *magnesium sulphate*, (2) *cadmium sulphate*, (3) *zinc sulphate*, to each of which you have added about $\frac{1}{10}$ or $\frac{1}{12}$ of its volume of concentrated sulphuric acid.

The arrangement of the experiment is shewn in fig. 25. Notice that a greyish white solid forms on the negative electrode in each case; these deposits are (1) *magnesium*, (2) *cadmium*, (3) *zinc*. Compare the result with that obtained in the electrolysis of hydrochloric acid (*Exp.* 15, *Chap. XII.*).

Fig. 25.

**Exp. 3.** Place the end of a piece of *magnesium* ribbon or wire in the flame of a Bunsen-lamp; the magnesium is rapidly burnt to magnesia with production of brilliant white light. Heat some *zinc* in an iron spoon over a blowpipe until it is quite melted; then direct a fairly rapid stream of *oxygen* from a gasholder over the molten metal, keeping the spoon in the blowpipe flame. The zinc is burnt to zinc oxide. Repeat this experiment with *cadmium* in place of zinc; the cadmium is burnt to cadmium oxide. The reaction in each case is;— $M + O = MO$ where $M = Mg$, Zn, or Cd.

**Exp. 4.** Shake up a little *magnesium oxide, zinc oxide*, and *cadmium oxide*, with *water* for a few minutes; then divide each into two parts and add to these red and blue litmus, respectively. The red litmus is turned blue, but not deep blue, in the case of the magnesium oxide; the other oxides do not change the colour of either the red or the blue litmus.

Magnesium oxide is then slightly alkaline.

Add some of each oxide to a little warm *sulphuric acid*, continuing to add the oxide until some remains undissolved. Filter, and evaporate each filtrate to the crystallising point; Collect each crop of crystals, dry them, and purify them by recrystallisation from water. Prove that the crystals obtained by using magnesium oxide are (1) a compound of magnesium, (2) a sulphate; that the crystals obtained by using zinc oxide are (1) a compound of zinc, (2) a sulphate; and that the crystals obtained by using cadmium oxide are (1) a compound of cadmium, (2) a sulphate.

*The production of a yellow pp. of cadmium sulphide* (CdS), *insoluble in warm ammonium sulphide, when sulphuretted hydrogen is passed into an acidulated solution, is a test for the presence of cadmium in that solution.*

The three oxides experimented with are therefore basic oxides; they interact with acids to produce salts. The change which occurred in the formation of the sulphates of magnesium, zinc, and cadmium, is represented thus;

$MO + H_2SO_4Aq = MSO_4Aq + H_2O$, where $M = Mg$, $Zn$, or $Cd$.

The crystals of the three sulphates have the compositions $MSO_4.7H_2O$ when $M = Mg$ or $Zn$, and $3CdSO_4.8H_2O$.

**Exp. 5.** To aqueous solutions of *magnesium sulphate, zinc sulphate*, and *cadmium sulphate*, add *caustic potash*; collect the white pps. which are produced, on filters, and wash each with cold water until the washings are neutral to litmus and free from sulphates. Dry the pps. at 100°. These pps. are hydroxides having the composition $MO_2H_2$ where $M = Mg$, $Zn$, or $Cd$. Prove that zinc hydroxide is soluble in excess of potash, but the others are insoluble. Place a part of each dried pp. in a large, clean, dry, test tube, and heat the lower part of the tube. Water is given off in each case. The hydroxides are changed to oxides and water; thus, $MO_2H_2 = MO + H_2O$.

**Exp. 6.** Repeat last exp. but use *sodium carbonate* in place of caustic potash; pps. of magnesium, zinc, and cadmium, carbonates are produced, with evolution of carbon dioxide in each case; boil the liquids holding the pps. in suspension for some time; then filter, wash the pps. with hot water until the washings are free from sulphates and carbonates, and dry the pps. at 100°. You thus obtain the carbonates $MCO_3$, combined with more or less of the hydroxides $MO_2H_2$, where $M = Mg$, $Zn$, or $Cd$. Place a small quantity of each carbonate in a crucible and heat strongly over a blowpipe flame until a very little of the solid taken out and added to a few drops of dilute hydrochloric acid in a test tube ceases to evolve any gas as it dissolves in the acid. The residues are oxides $MO$: the change which has occurred is $MCO_3 = MO + CO_2$.

These experiments afford reasons for placing the three elements magnesium, zinc, and cadmium in the same class; they also shew that there are slight differences in the chemical behaviours of the three elements under similar conditions.

*Reference to* "ELEMENTARY CHEMISTRY." Chap. XIX.

# CHAPTER XIV.

## CLASSIFICATION OF ELEMENTS (*continued*).

The elements nitrogen, phosphorus, arsenic, antimony, and bismuth are placed in the same class.

Compare the prominent physical properties of the five elements.

*Nitrogen*: a colourless, odourless, gas.

*Phosphorus*: a soft, waxy, crystalline, solid; without metallic lustre; low spec. gravity.

*Arsenic* and *antimony*: lustrous, grey, brittle, crystalline, solids.

*Bismuth*: heavy, brittle, lustrous, crystalline, solid.

Nitrogen, phosphorus, arsenic, and antimony form compounds with hydrogen, $MH_3$.

**Exp. 1.** Arrange an apparatus as shewn in fig. 26. $A$ is a dry flask containing a mixture of about 20 grams powdered *ammonium chloride* ($NH_4Cl$) with an equal weight of *lime* ($CaO$); this mixture is covered with a layer of lime; $B$ is a tower containing loosely packed pieces of lime; $c$ is a *dry* bell-jar. Have ready another dry jar, and a large dry flask to which you have fitted a good caoutchouc cork carrying a piece of rather wide glass tube narrowed at the end which goes into the flask, with a caoutchouc tube attached and a pinchcock on the caoutchouc tube (Fig. 27). Heat the flask $A$ on a sand-tray; ammonia, $NH_3$, comes off, is dried by passing through the lime in $B$, and collects in the jar $c$. The reaction is $2NH_4Cl + CaO = 2NH_3 + CaCl_2 + H_2O$; the layer of lime absorbs the water produced and prevents it from trickling down and so probably breaking the flask.

*As ammonia is much lighter than air, it is easily collected by upward displacement.*

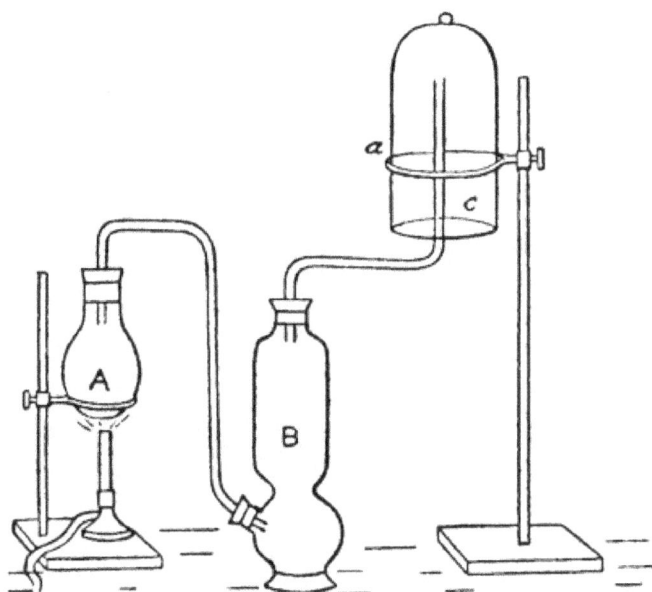

Fig. 26.

The jar *c* is full of ammonia when a piece of moist red litmus paper held about the place marked *a* (Fig. 26) is at once turned blue. When the jar is full of the gas, cover its mouth with a well greased glass plate, and place it aside, *mouth downwards.* Fill another jar, and the dry flask, with ammonia. When the flask is full pass the stream of ammonia through the caoutchouc and glass tubes, fit the cork tightly into the flask, and at once close the pinchcock.

**Exp. 2.** Arrange the flask, full of *ammonia*, as shewn in fig. 27. The beaker contains *water* coloured red by *litmus* solution. Open the pinchcock; pour a little ether over the outside of the flask to cool the contents. The water rises in the tube; as soon as the water reaches the flask the ammonia is very rapidly dissolved in it, the pressure inside the flask is rapidly diminished, more water is forced up the tube, more ammonia is at once dissolved, and so on until the whole of the ammonia in the flask has been dissolved.

The red colour of the litmus is changed to blue.

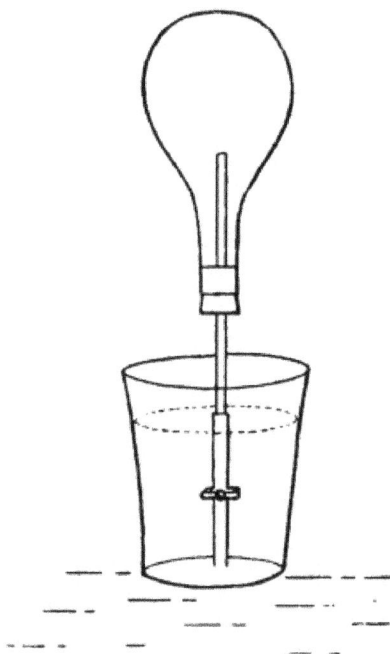

Fig. 27.

Ammonia therefore is very soluble in water, and this solution acts towards litmus as alkalis do.

**Exp. 3.** To one of the jars of *ammonia* add a few drops of concentrated *hydrochloric acid*; white fumes fill the jar, and settle down as a fine white solid on the sides.

This white solid is ammonium chloride; the reaction is

$$NH_3 + HCl = NH_4Cl.$$

**Exp. 4.** Bring a lighted taper to the mouth of the other jar containing *ammonia*; then pass the taper into the jar. The gas does not burn, nor does it support combustion.

**Exp. 5.** Put a little concentrated *ammonia solution* in a small flask, gently warm the flask, then place it on the table, pass a fairly rapid stream of *oxygen* through the liquid, and bring a lighted taper to the mouth of the flask. Under these conditions the ammonia burns with a greenish flame; the products are nitrogen and water, and sometimes a little nitrogen oxide.

The compound of phosphorus and hydrogen, phosphuretted hydrogen or phosphine, $PH_3$, can be obtained by a reaction similar to that whereby ammonia was made in Exp. 1 ; thus $PH_4I + KOHAq = PH_3 + KIAq + H_2O$. But the salt $PH_4I$ (phosphonium iodide) is rather difficult to prepare.

**Exp. 6.**  Arrange a small flask, about 100 to 150 c.c. capacity, as shewn in fig. 28. The tube $a$ is rather wide, it passes a little way through the cork, and carries a piece of caoutchouc tubing fitted with a pinchcock. The basin $c$

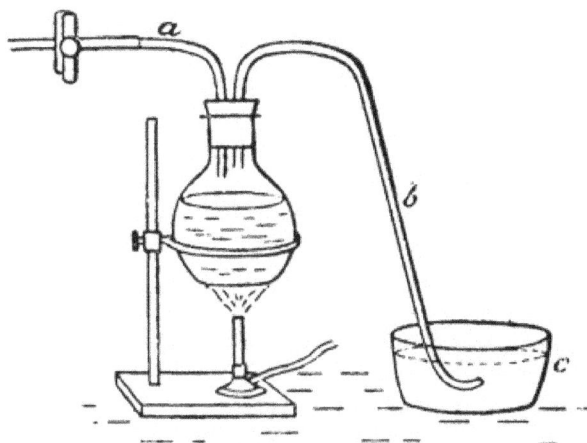

Fig. 28.

contains water; the end of the delivery tube $b$ remains beneath the surface of this water. Nearly fill the flask with fairly concentrated *potash* solution; put about 6 or 8 *small* pieces of *phosphorus* in the flask; fit the cork into the flask; connect $a$ with the gas-tap and pass a stream of coal-gas through the whole apparatus for some minutes. Now stop the gas-stream; at once close the pinchcock near the end of $a$, and gently warm the contents of the flask. After a little, bubbles of gas escape through the water in $c$; each bubble takes fire as it comes into the air and burns to a ring of white smoke. This smoke consists of phosphorus pentoxide; the reaction may be represented thus ; $2PH_3 + 8O = P_2O_5 + 3H_2O$.

The change which occurs in the flask may be thus represented in an equation ;—

$$3KOHAq + 4P + 3H_2O = PH_3 + 3KH_2PO_2Aq.$$

The gas which is produced in the experiment contains a little of the hydride $P_2H_4$; it is this body which takes fire in the air; pure $PH_3$ is not inflammable at ordinary temperatures.

**Exp. 7.**    Arrange an apparatus as shewn in fig. 29.    Put some pure granulated *zinc* and water in the flask $A$; $B$ is a tube containing calcium chloride which serves to dry the gas passing through it; $c$ is a tube of hard glass narrowed to a

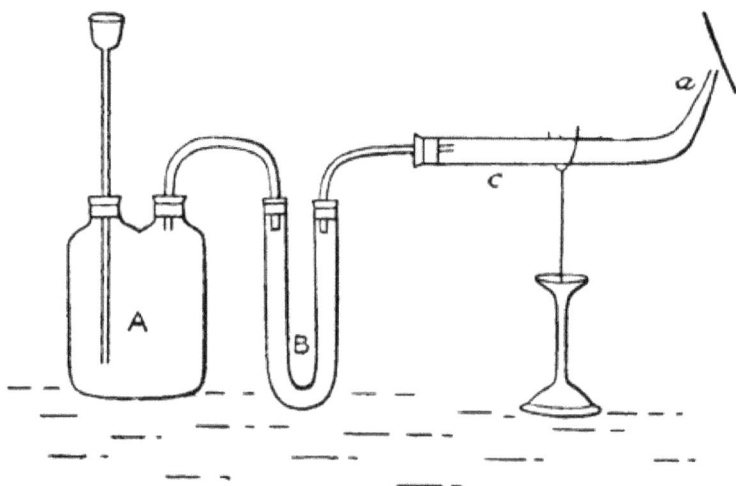

Fig. 29.

fine opening at the end $a$.    Pour a little pure dilute *sulphuric acid* through the funnel tube into $A$; allow the escaping hydrogen to drive out all the air from the apparatus (*s. Exp. 2, Chap. VI.*), and then light the gas at $a$.    *Place the apparatus in the draught place\*, and then pour a very small quantity* of an aqueous solution of *arsenious oxide* ($As_2O_3$) through the funnel tube into $A$.    The escaping hydrogen is now mixed with arsenious hydride, $AsH_3$.    The reaction may be thus formulated, $As_2O_3 + 12H = 2AsH_3 + 3H_2O.$

In order to bring about this reaction the hydrogen must be prepared in contact with the arsenious oxide solution (*s. ELEMENTARY CHEMISTRY, Chap. XI. par.* 175).

* $AsH_3$ is extremely poisonous.

Note the change in the colour and appearance of the flame at $a$, and the garlic-like odour produced; the arsenious hydride is being burnt to oxide and water, thus

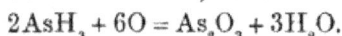

$$2AsH_3 + 6O = As_2O_3 + 3H_2O.$$

Bring a porcelain crucible-lid *over* the flame; the arsenious oxide is deposited on the lid as a white, nearly invisible, film.

Now bring a porcelain crucible-lid *into* the flame as shewn in fig. 28. The supply of oxygen is thus limited, and at the same time the flame is cooled; the change which now proceeds is chiefly this;—$2AsH_3 + 3O = 2As + 3H_2O$. The arsenic is deposited on the cool surface of the lid.

Now heat the tube through which the mixture of hydrogen and arsenious hydride is passing at about the point $c$; a shining metal-like deposit of arsenic is slowly formed nearer the open end of the tube than the heated part. This deposit is soluble in a solution of bleaching powder. The gas $AsH_3$ is decomposed by heat into its elements; the arsenic is deposited on the cold parts of the tube, and the hydrogen passes on.

Turn the tube $c$ so that the open end points downwards; let this end dip into an aqueous solution of *silver nitrate* ($AgNO_3$) in a test tube; after the gas has passed through this solution for a little, remove the tube, and pour the liquid through a filter; to the filtrate add *one or two drops* of *dilute* ammonia solution; a pale yellow pp. of silver arsenite is produced; add a little more ammonia, the pp. dissolves. Wash the black solid on the filter several times with hot water, then dissolve it in warm *nitric acid* and prove that the solution contains silver. The interaction which has occurred is represented thus;—$12AgNO_3Aq + xAgNO_3Aq + 2AsH_3 + 3H_2O = xAgNO_3Aq + As_2O_3Aq + 12HNO_3Aq + 12Ag$; the silver nitrate is *reduced* by the arsenious hydride. The addition of ammonia after filtering neutralises the nitric acid formed in the reaction, the arsenious oxide then interacts with the silver nitrate remaining in solution to produce silver arsenite; thus,

$$6AgNO_3Aq + As_2O_3Aq + 3H_2O = 2Ag_3AsO_3 + 6HNO_3Aq.$$

**Exp. 8.**  Arrange another apparatus as shewn in fig. 29, and conduct a series of experiments as described under Exp. 7, but use a solution of an *antimony compound* (an aqueous solution of *tartar emetic* $KSbC_4H_4O_7$ will do well) in place of arsenious oxide.

Note (1) the colour of the flame of the mixture of hydrogen and antimony hydride ($SbH_3$);

(2) the appearance of the deposit of antimony on the porcelain lid ;

(3) the position and appearance of the deposit of antimony on the tube, and its insolubility in bleaching powder solution ;

(4) the fact that after passing the gas through silver nitrate solution, and filtering, no antimony remains in solution ;—prove this by adding hydrochloric acid to the filtrate, filtering off the pp. of silver chloride so produced, and passing a little sulphuretted hydrogen into the filtrate ; no pp. is formed ; had antimony been present orange-red $Sb_2S_3$ would have been precipitated.

The interaction between antimony hydride and silver nitrate in solution may be represented thus ;—

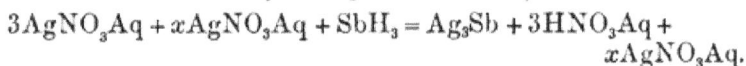

$$3AgNO_3Aq + xAgNO_3Aq + SbH_3 = Ag_3Sb + 3HNO_3Aq + xAgNO_3Aq.$$

Exps. 1 to 8 exhibit the methods of preparation and some of the important properties of the hydrides $MH_3$ when $M = N$, P, As, Sb.

The five elements, nitrogen, phosphorus, arsenic, antimony, and bismuth form each more than one oxide; we shall prepare some of the oxides having the composition $M_2O_3$ and examine their properties.

The oxide $N_2O_3$ is obtained by partially deoxidising nitric acid ($HNO_3$).

**Exp. 9.**   Place about 5 grams *arsenious oxide* in a test

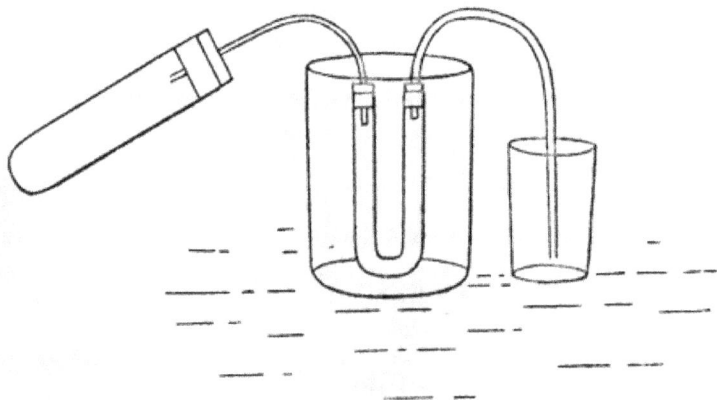

Fig. 30.

tube arranged with a cork and exit tube passing into a U tube as shewn in fig. 30. Add about 20 c.c. of a mixture of equal volumes concentrated *nitric acid* and water. Surround the U tube with a mixture of ice and salt and gently heat the test tube. A blue-coloured liquid collects in the U tube; this is impure $N_2O_3$:—

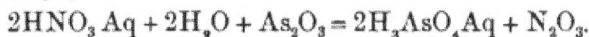

$$2HNO_3 Aq + 2H_2O + As_2O_3 = 2H_3AsO_4Aq + N_2O_3.$$

Put about 20 c.c. of water in a beaker surrounded by ice, and after a little time add 2 or 3 c.c. of the blue liquid from the U tube. Now prove (1) that the solution has an acid reaction to litmus paper, (2) that it contains *nitrous acid.*

$$(N_2O_3 + H_2O = 2HNO_2Aq).$$

*To test for nitrous acid or nitrites, add a few drops of an aqueous solution of potassium iodide* (KI) *and a very little starch paste* (*prepared by boiling a very small quantity of starch with water and allowing to cool*) *to the acidulated liquid; a blue colour, due to "iodide of starch," is produced.* The reaction is $2KNO_2Aq + 2KIAq + 4HCl = 4KClAq + 2H_2O + 2NO + I_2.$

**Exp. 10.** The oxide $P_2O_3$ is prepared by oxidising phosphorus in a limited supply of air. This oxide interacts with water to form a solution of phosphorous acid, $H_3PO_3Aq.$

The apparatus represented in fig. 31 consists of a piece of

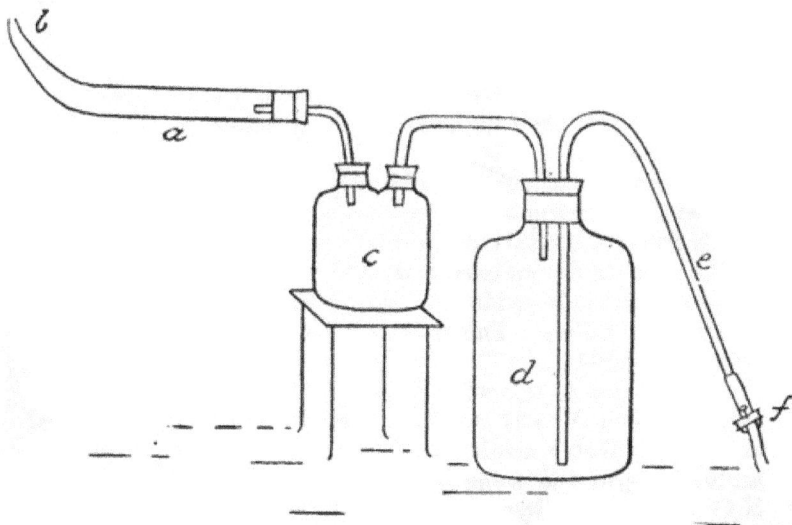

Fig. 31.

hard glass tubing, $a$, narrowed at one end $b$, and fitted at the other end with a cork carrying a tube which passes into a dry two-necked bottle, $c$, this bottle communicates with a large bottle or jar, $d$, as shewn in the figure ; the tube $e$ acts as a syphon, it is furnished with a stopcock $f$.

Fill the bottle $d$ with water ; arrange the apparatus as shewn ; then cut three or four small pieces of *phosphorus* from the middle of a stick of phosphorus and place them in the tube $a$. Partly open the stopcock $f$; a slow current of *air* is sucked through the entire apparatus. Now gently heat the phosphorus ; it is burnt in the limited air-supply ; the chief product of the burning is $P_2O_3$, which collects, as a white solid, on the cool parts of $a$ and in the bottle $c$. Stop the process before the whole of the phosphorus is burnt ; prove that the white solid obtained dissolves in water and that this solution contains an acid.

The oxides $As_2O_3$, $Sb_2O_3$, and $Bi_2O_3$, may be prepared by burning the elements As, Sb, or Bi, in air or oxygen. The oxides $As_2O_3$ and $Sb_2O_3$ are slightly acidic, and $Sb_2O_3$ is also feebly basic ; $Bi_2O_3$ is basic.

**Exp. 11.** Prove (1) that *arsenious oxide* ($As_2O_3$) is very slightly soluble in water and that the solution scarcely affects the colour of blue litmus paper ; (2) that *antimonious oxide* ($Sb_2O_3$) is almost insoluble in water ; (3) that *bismuthous oxide* ($Bi_2O_3$) is quite insoluble in water. Dissolve a little $Sb_2O_3$ and $Bi_2O_3$, separately, in concentrated *sulphuric acid*, evaporate, and allow to crystallise ; in each case crystals are obtained ; these crystals have the composition $M_2 3SO_4$ where $M = Sb$ or Bi.

To a little *antimonious oxide* add some concentrated *nitric acid*, and heat ; the oxide does not dissolve ; the white solid which remains is a mixture of $Sb_2O_4$ and $Sb_2O_5$.

To a little *bismuthous oxide* add some concentrated *nitric acid*, and heat ; the oxide dissolves ; evaporate the solution, and allow to crystallise. The crystals which separate are bismuth nitrate, $Bi3NO_3$.

The oxides $M_2O_5$, when $M = N$, P, As, or Sb, are acidic oxides : when $M = Bi$ the oxide is a peroxide possessed of exceedingly feeble acidic functions ; this oxide interacts with acids to form the same salts as are obtained from $Bi_2O_3$. $N_2O_5$ is prepared by removing water from $HNO_3$ ; $P_2O_5$ by burning P in excess of air or oxygen ; $As_2O_5$ and $Sb_2O_5$ by evolving oxygen in contact with the elements, or the oxides $M_2O_3$, by the decomposition of concentrated nitric acid; and $Bi_2O_5$ by evolving oxygen in contact with $Bi_2O_3$ in presence of much potash or soda.

**Exp. 12.**   Burn a small piece of *phosphorus* in a large bottle full of *air*: dissolve the white solid which is formed ($P_2O_5$) in water and prove that the solution turns blue litmus red. (*s. also Exps.* 15 *and* 16.)

To a small quantity of *arsenious oxide* add some concentrated *nitric acid*, heat for some time, then pour off the remaining acid, add more concentrated nitric acid and heat; then evaporate in the draught place just to dryness, and raise to a low red heat.   Collect the $As_2O_5$ which remains and then dissolve it in water; prove that this solution contains an acid, and that this is arsenic acid.

*To test for arsenic acid* ($H_3AsO_4$) *add silver nitrate, when a red pp. of silver arsenate,* $Ag_3AsO_4$, *forms; this pp. is soluble in hot concentrated nitric acid, and is reprecipitated on concentration and cooling.*

Arrange an apparatus as shewn in fig. 32.   Into the flask

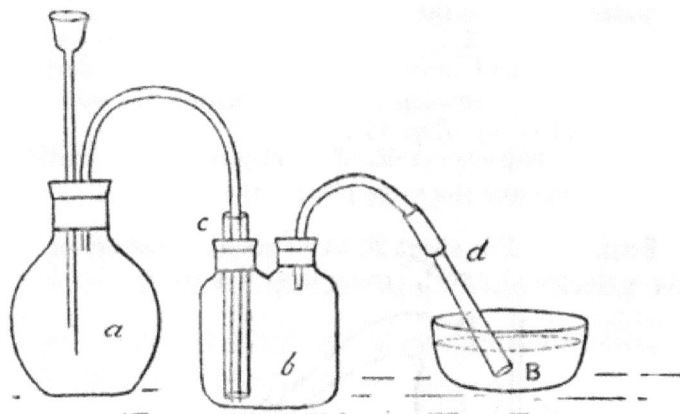

Fig. 32.

*a* put several lumps of *manganese peroxide* ($MnO_2$); put dilute *potash* solution in the bottle *b*; the tube *c* is wider than the delivery tube from *a*; *d* is a tube about 10 mm. wide at the

open end and narrowed at the other end so as to be easily and
firmly connected, by caoutchouc, with the delivery tube from *b*.
Put a little *bismuthous oxide* and a quantity of a *completely saturated* aqueous solution of *potash* in the basin *B*. Arrange the
whole apparatus so that the flask *a* can be heated by a Bunsen-
lamp; pour some concentrated hydrochloric acid into *a*, and
gently warm the flask. Chlorine is evolved (*s. Chap. XII. Exp.* 1);
any traces of acid carried over from *a* are absorbed in *b*.
Continue passing *chlorine* into the basin *B* until the liquid
smells distinctly of chlorine; then remove *a* by drawing the
delivery tube out of the wider tube *c*, and set the flask aside.
Heat the contents of *B* to boiling; then add more potash, and
again pass in chlorine until there seems to be no further
change in the appearance of the solid matter in the basin.
In this reaction potassium hypochlorite (KClO) is produced
and is again decomposed (KClOAq heated in presence of an
alkali gives KCl + O), and the oxygen thus evolved oxidises
the $Bi_2O_3$ to $Bi_2O_5$. Pour off the liquid from the reddish solid
in *B*, wash the solid repeatedly with boiling water, then with a
little dilute warm nitric acid to remove the last traces of
potash which are retained by the $Bi_2O_5$, and then again with
hot water to remove the nitric acid.

Dissolve the $Bi_2O_5$ thus produced in warm concentrated
sulphuric acid and prove that oxygen is evolved during the
solution. On evaporation and crystallisation, the salt $Bi_23SO_4$
separates out (*comp. Exp.* 11).

The most important acids of the elements we are considering
are nitric acid and the three phosphoric acids.

**Exp. 13.** Put about 20 to 30 grams of *potassium nitrate*
into a stoppered retort, arranged with a receiver as shewn in

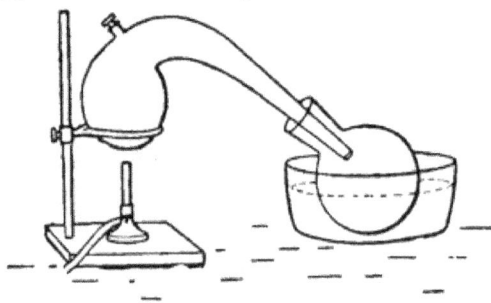

Fig. 33.

fig. 33. Pour enough concentrated *sulphuric acid* into the retort to completely cover the crystals of potassium nitrate. Place the receiver in cold water, and allow a stream of water to flow over its outer surface. On warming the contents of the retort, nitric acid is formed, and is condensed to a yellowish liquid in the receiver;—$KNO_3 + H_2SO_4 = KHSO_4 + HNO_3$.

Part of the nitric acid is decomposed by the action of the hot sulphuric acid, giving water, oxygen, and brown fumes of nitrogen tetroxide, $N_2O_4$. Boil a little nitric acid in a small flask arranged with a cork and exit tube dipping under water, and prove that oxygen is evolved ; notice also the brown gas which is produced. The change may be represented thus

$$2HNO_3 = H_2O + O + N_2O_4.$$

Nitric acid then ought to act as an oxidising agent. To prove that it does thus act, perform the next Exp.

**Exp. 14.** Pour a little of the *nitric acid* you have prepared, drop by drop, on to some powdered *antimony* in a basin ; the antimony is oxidised to white antimony oxide ($Sb_2O_4$).

We shall now take advantage of the oxidising power of nitric acid in order to prepare phosphoric acid from phosphorus.

**Exp. 15.** Place a few small pieces of *phosphorus* in a basin, add some concentrated *nitric acid*, and heat so long as a brown gas (chiefly $N_2O_4$) is evolved ; then add a little more nitric acid ; and repeat this treatment until the phosphorus has been wholly dissolved. Heat the basin over a low flame in the draught place until the liquid becomes thick and syrupy ; add a little water, and heat again. The unchanged nitric acid is thus removed, and a concentrated solution of orthophosphoric acid ($H_3PO_4$) remains

$$2P + 10HNO_3Aq = 2H_3PO_4Aq + 2H_2O + 5N_2O_4.$$

*Neutralise a small quantity of this solution by ammonia, and then add silver nitrate solution ; a pale yellow pp. of silver orthophosphate* ($Ag_3PO_4$) *is produced.*

Now dissolve a little phosphorus pentoxide ($P_2O_5$) in water ; boil for some time, and prove, by the foregoing test, that the solution contains orthophosphoric acid. The change is

$$P_2O_5 + 3H_2O + Aq = 2H_3PO_4Aq.$$

Boil the rest of the solution of orthophosphoric acid which you prepared to dryness in a platinum dish, and continue to heat the residue over a Bunsen-lamp for some little time. On cooling, you obtain a solid glass-like mass. This is meta-phosphoric acid $HPO_3$ ($H_3PO_4 - H_2O = HPO_3$). Dissolve this solid in water, and test the solution; (1) by *neutralising by ammonia and adding silver nitrate solution, a white pp. of silver metaphosphate* ($AgPO_3$) *is formed*; (2) *by adding an albumen-containing liquid (a very little white of egg shaken up with much water), the albumen is coagulated and precipitated*.

Now dissolve a little phosphorus pentoxide in a very little ice-cold water, and prove that this solution contains meta-phosphoric acid. The change is

$$P_2O_5 + H_2O + \text{a little } Aq = 2HPO_3Aq.$$

There is another phosphoric acid, called pyrophosphoric, $H_4P_2O_7$; this may be obtained by heating orthophosphoric acid to about $250°$ until water is no longer removed ($2H_3PO_4 - H_2O = H_4P_2O_7$), but it is more easily prepared from lead pyrophosphate, which may be readily obtained from sodium pyrophosphate, which salt is itself produced by heating ordinary sodium orthophosphate. This process is repre-sentative of a commonly employed method for preparing acids from their salts.

**Exp. 16.** Prove, by the silver nitrate test (Exp. 15), that ordinary *sodium phosphate* is a salt of orthophosphoric acid. Now heat some sodium phosphate to redness in a platinum dish as long as there is any apparent change. Allow the solid residue to cool, dissolve in *cold* water and apply the following tests to small quantities of the solution :—

(1)   *silver nitrate*; a white pp. is produced.

(2)   *albumen* is not coagulated, after the solution has been acidified by acetic acid.

The salt you have prepared is sodium pyrophosphate; the change from orthophosphate may be thus represented;
$2Na_2HPO_4 - H_2O = Na_4P_2O_7$.

Prepare lead pyrophosphate, $Pb_2P_2O_7$, by adding an aqueous solution of *lead nitrate* to the remainder of the solution of *sodium pyrophosphate*; collecting the white pp. which forms, and washing it with cold water until the washings are free from nitrates :—

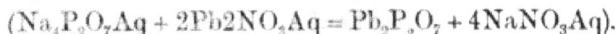

$$(Na_4P_2O_7Aq + 2Pb2NO_3Aq = Pb_2P_2O_7 + 4NaNO_3Aq).$$

Now suspend the greater part of the lead pyrophosphate in cold water, pass a stream of washed *sulphuretted hydrogen* ($H_2S$) through the liquid until it smells decidedly of the gas; then add the remainder of the lead pyrophosphate, in order to remove the excess of $H_2S$ which is present, and filter off the pp. of PbS and excess of $Pb_2P_2O_7$. The changes which have occurred may be represented thus :—

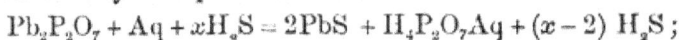

$$Pb_2P_2O_7 + Aq + xH_2S = 2PbS + H_4P_2O_7Aq + (x-2) H_2S ;$$

then the remaining $H_2S$ is decomposed by the $Pb_2P_2O_7$ added, and the excess of $Pb_2P_2O_7$ remains unchanged.

Neutralise by ammonia the aqueous solution of pyrophosphoric acid which you have thus prepared ; and prove that it then gives the same reactions as the solution of sodium pyrophosphate already tested. Boil a portion of the solution and prove that it now contains orthophosphoric acid ($H_4P_2O_7Aq + H_2O = 2H_3PO_4Aq$).

The foregoing exps. (13 to 16) have taught us that the oxide $N_2O_3$ interacts with water to form one acid only ; and that the oxide $P_2O_5$ interacts with water to form three distinct acids : of these, orthophosphoric ($H_3PO_4$) is the most stable, as regards the action of heat, in aqueous solution, but metaphosphoric ($HPO_3$) is the most stable, as regards the action of heat, when in the solid state. (For some account of the acids obtained from the oxides of arsenic and antimony, see ELEMENTARY CHEMISTRY, Chap. XI. pars. 214—217.)

We shall now perform a few experiments illustrative of the preparation and properties of some of the sulphides of the elements we are considering.

**Exp. 17.** Through solutions (i) of *arsenious oxide* in very dilute hydrochloric acid, (ii) of *tartar emetic* in very dilute hydrochloric acid, (iii) of *bismuth chloride* in very dilute hydrochloric acid, pass *sulphuretted hydrogen* gas ; collect the pps. which form and wash each two or three times with water. The composition of these pps. is represented by the formula

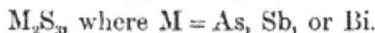

$$M_2S_3, \text{ where } M = As, \text{ Sb, or Bi.}$$

To separate small quantities of each pp. add (i) *ammonium sulphide* solution, (ii) *potash* solution, and warm ; the sulphides of arsenic and antimony dissolve in each case, the sulphide of bismuth remains unchanged.

To the solutions in ammonium sulphide add a good deal of *alcohol* ; white crystals are pptd. These crystals are ammonium thioarsenite, $(NH_4)_3AsS_3$, and thioantimonite $(NH_4)\ SbS_2$, respectively. The changes which occur when $As_2S_3$ and $Sb_2S_3$ severally interact with ammonium sulphide may be represented as follows :—

$$As_2S_3 + 3\,(NH_4)_2S = 2\,(NH_4)_3AsS_3,$$
$$Sb_2S_3 + (NH_4)_2S = 2\,(NH_4)SbS_2.$$

The sulphides of arsenic and antimony, $M_2S_3$, are therefore acidic, i.e. they interact with sulphides of very positive elements, or with sulphides of the positive compound radicle $NH_4$, to form salts. The corresponding sulphide of bismuth, $Bi_2S_3$, is not acidic. Compare the properties of the oxides $M_2O_3$ with those of the sulphides $M_2S_3$ where M = As, Sb, or Bi.

The foregoing experiments shew that the elements nitrogen, phosphorus, arsenic, antimony, and bismuth are analogous in their chemical properties; and that as the combining weights of the elements increase the elements become more metallic and less negative.

*Reference to* ELEMENTARY CHEMISTRY, Chap. XI. pars. 206 —225.

# CHAPTER XV.

CLASSIFICATION OF ELEMENTS (*continued*).

THE elements chromium, manganese, and iron, are placed in the same class.

These elements form hydrated oxides $M_2O_3 . 3H_2O$ where $M = Cr$, Mn, or Fe.

**Exp. 1.** To solutions of (i) *ferric chloride* ($Fe_2Cl_6$), and (ii) *chromic chloride* ($Cr_2Cl_6$), add a slight excess of *ammonia* solution. Collect the pps. which form; wash them repeatedly with hot water; and prove that a small quantity of each dissolves readily in hydrochloric acid; set these solutions evaporating in the draught cupboard, (*see next Exp.*). Now place the rest of each pp. in a small basin and heat these at 100° so long as water comes off. The solids have now the composition $M_2O_3.3H_2O$ where $M = Fe$ or Cr. Heat these solids strongly over a Bunsen-lamp; water is removed and the oxides $M_2O_3$ remain. Prove that the oxides thus prepared are nearly insoluble in concentrated hot hydrochloric acid.

From these reactions we would expect that a solution of a manganic salt (say manganic chloride) would interact with ammonia to produce a pp. of the composition $Mn_2O_3.3H_2O$. But the salts of manganese corresponding to $Fe_2Cl_6$ and $Cr_2Cl_6$ are extremely unstable and can hardly be obtained even approximately pure.

**Exp. 2.** Add some *ammonium chloride* and then a slight excess of *ammonia* to an aqueous solution of *manganous chloride* ($MnCl_2$); an almost white pp. of manganous hydrate, $MnO.H_2O$, forms. Pour the contents of the tube into a large stoppered bottle, and shake up the pp. repeatedly, removing the stopper at intervals and blowing in fresh supplies of air. The pp. slowly turns brown; the brown solid eventually produced is hydrated manganic oxide, $Mn_2O_3.H_2O$. (Compare the composition of this compound with that of the compounds produced from ferric and chromic chlorides; $M_2O_3.3H_2O$.) The change which has proceeded in the bottle may be thus shewn ; $2MnO.H_2O + O$ (from air) $= Mn_2O_3.H_2O + H_2O$. Collect the brown solid on a filter, wash it well with warm water, dissolve a portion in warm concentrated hydrochloric acid, and set this solution evaporating in the draught place; keep the rest of the $Mn_2O_3.H_2O$ for the next Exp.

When the solutions in hydrochloric acid of the hydrated oxides $M_2O_3$ have been evaporated nearly to dryness, each yields a solid on cooling ; these solids are yellow-brown ferric chloride $Fe_2Cl_6$, greenish-violet chromic chloride $Cr_2Cl_6$, and slightly pink manganous chloride $MnCl_2.4H_2O$.

**Exp. 3.** To solutions of *ferric chloride* ($Fe_2Cl_6$) and *chromic chloride* ($Cr_2Cl_6$) add a solution of *caustic potash* until a pp. is formed in each case. These pps. are $Fe_2O_3.3H_2O$ and $Cr_2O_3.3H_2O$ respectively. Now add a considerable excess of potash to each pp., the $Cr_2O_3.3H_2O$ dissolves but the precipitated ferric hydrate remains unchanged. Now add a considerable quantity of potash to the pp. of $Mn_2O_3.H_2O$ remaining from Exp. 2 ; it remains unchanged. Hydrated chromic oxide therefore differs from the corresponding compounds of iron and manganese in being soluble in a solution of caustic potash (or soda). As solutions of oxides in caustic alkalis often contain salts the negative parts of which are composed of the oxides in question (e.g. $P_2O_5$ dissolves in KOHAq forming $P_2O_5.3K_2OAq$), it may be that the fact of the solubility of $Cr_2O_3.3H_2O$ in potash shews that $Cr_2O_3.3H_2O$ is slightly acidic in its functions. (*s. further Exps.* 7, 10, *and* 11 *in this Chap.*).

The hydrated ferrous and chromous oxides $MO.H_2O$ where $M = Cr$ or $Fe$ are very unstable; they quickly combine with oxygen and become $M_2O_3.3H_2O$. But it is possible to prepare $FeO.H_2O$ nearly free from ferric hydrate.

**Exp. 4.** Arrange an apparatus as shewn in fig. 34. *A* is a flask containing zinc and dilute sulphuric acid ; *B* and *C* are each short wide test tubes supported in stands ; each of these tubes is about half filled with distilled water, and a very little sulphuric acid is also poured into *C*; *a* and *b* represent pieces of caoutchouc tubing. *Hydrogen* is passed through the whole apparatus for some minutes; the corks are then loosened in

*B* and *C* (the hydrogen is allowed to continue passing through the tubes); the contents of each tube are boiled for 5 or 10 minutes, to remove the air dissolved in the water.   The corks

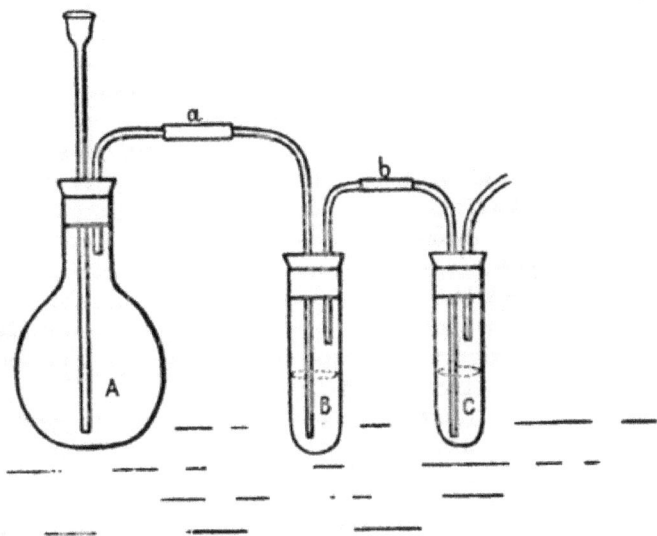

Fig. 34.

are now replaced in *B* and *C* and the passage of hydrogen is continued; when the water in *B* and *C* is cold, the corks are removed for an instant, a piece of solid *potash* is dropped into *B*, and a small piece of granulated *zinc* and a clear crystal of *ferrous sulphate* ($FeSO_4.7H_2O$) into *C*, and the corks are quickly replaced.   After a minute or two the tube *B* is inverted; the hydrogen then forces the potash solution from *B* into *C*, and the potash and ferrous sulphate interact to produce white ferrous hydrate and potassium sulphate; thus $2KOHAq + FeSO_4Aq = FeO.H_2O + K_2SO_4Aq$.   Now remove the cork from *C*, and pour the contents into a vessel where they are freely exposed to air; the colour of the pp. soon changes to greenish, then to greenish-brown, and then to brown ($Fe_2O_3.3H_2O$ is eventually produced).

The elements we are considering form oxides of the composition $M_3O_4$.

**Exp. 5.**   Dissolve about 20 grams of *ferrous sulphate* in cold water; convert about $\frac{2}{3}$ of this into a solution of ferric sulphate by adding some *sulphuric acid*, heating to boiling, and adding concentrated *nitric acid*, drop by drop, so long as a change of colour is produced. Allow the solution of ferric sulphate to cool; then mix it with the ferrous sulphate solution; add a slight excess of ammonia solution; boil for some time; allow the pp. of $Fe_3O_4$ which is produced to settle, wash it thoroughly with hot water by decantation, transfer it to a porcelain dish, and dry it at about 60°.

Prove that the $Fe_3O_4$ you have prepared dissolves in HClAq with production of both ferrous and ferric chloride; making use of the facts (1) *that potassium sulphocyanide solution* (KCNSAq) *gives a deep red colour* [$Fe(CNS)_3Aq$] *with ferric salts*, (2) *that potassium ferricyanide solution* ($K_3FeCy_6Aq$) *gives a pp. of prussian blue* ($Fe_7Cy_{18}$) *with ferrous salts*. Compare this reaction with that which occurs when HClAq and $Mn_3O_4$ interact; prove that manganous chloride ($MnCl_2$) is produced, and chlorine is evolved, in this case.

Contrast the reactions between (1) dilute nitric acid and $Fe_3O_4$, (2) dilute nitric acid and $Mn_3O_4$; prove that in the former case the $Fe_3O_4$ dissolves and that the liquid contains both ferrous and ferric nitrate, and that in the latter a brown solid remains undissolved—this solid is $MnO_2$—and the solution contains manganous nitrate.

*Prove the presence of a manganous salt in the liquid filtered from the insoluble* $MnO_2$ *by adding* $NH_3Aq$ *in slight excess and then* $NH_4HSAq$; *a buff coloured pp. of MnS is produced.*

Besides the oxides examined, manganese forms a peroxide $MnO_2$, and chromium a peroxide $CrO_3$.

**Exp. 6.** To an aqueous solution of a *manganous salt*, say $MnSO_4$, add a saturated solution of *sodium hypochlorite* (NaClO), and a considerable quantity of concentrated *soda* solution; heat until the pp. which forms is very dark brown and apparently homogeneous; collect and wash this pp. It is manganese dioxide or peroxide ($MnO_2$).

Recall the results of Exp. 2 wherein hydrated $Mn_2O_3$ was obtained by the action of air upon precipitated MnO; in the present Exp. the NaOHAq precipitates hydrated MnO, and at the same time this is oxidised to $MnO_2$ by the oxygen produced by the decomposition of the NaClO; (NaClOAq heated in an alkaline solution gives NaClAq + O).

**Exp. 7.** To about 200 c.c. of a cold saturated aqueous solution of *potassium dichromate* ($K_2Cr_2O_7$) add, slowly and with constant stirring, about 300 c.c. concentrated *sulphuric acid*, and allow the mixture to cool. Red crystals of $CrO_3$ separate out. $(K_2Cr_2O_7Aq + H_2SO_4 = K_2SO_4Aq + H_2O + 2CrO_3)$. Place a little *glass wool* in the neck of a funnel, and pour the liquid through this funnel; when the liquid has drained away from the crystals, remove the latter, by the aid of a platinum or glass spatula, to a clean dry porous tile, and allow them to drain there for a short time; then move the crystals to a fresh part of the porous tile and pour a very little *cold water* over them; when the water has soaked into the tile, repeat this treatment with another very small quantity *of cold water*. Now dissolve the crystals in a beaker in as small a quantity of hot water as possible, evaporate the solution and allow it to crystallize. Again collect the crystals, and dry them on a dry porous tile. In this way you obtain chromium trioxide or peroxide $(CrO_3)$ nearly freed from adhering sulphuric acid.

The oxides $MnO_2$ and $CrO_3$ readily part with a portion of their oxygen, $MnO_2$ being reduced to $MnO$, and $CrO_3$ to $Cr_2O_3$.

**Exp. 8.** To a solution of *oxalic acid* $(H_2C_2O_4)$ add some *manganese dioxide* $(MnO_2)$, and a little *sulphuric acid*, and heat nearly to boiling. The oxalic acid is oxidised to water and carbon dioxide; and the MnO produced is dissolved in the sulphuric acid forming $MnSO_4Aq$. Prove that $CO_2$ is evolved, and that the liquid gives the ordinary reactions of a manganous salt (*s. Exp.* 5).

**Exp. 9.** Place a little of the *chromium trioxide* $(CrO_3)$ you prepared in Exp. 7 in a basin and allow a little *alcohol* $(C_2H_6O)$ to slowly trickle on to it: the alcohol is at once oxidised to aldehyde, and green $Cr_2O_3$ remains; $(3C_2H_6O + 2CrO_3 = Cr_2O_3 + 3C_2H_4O + 3H_2O)$. Much heat is produced in this process and a portion of the alcohol is usually inflamed.

Chromium trioxide (or chromium peroxide) $CrO_3$ is a markedly acidic oxide.

**Exp. 10.** Dissolve the remainder of the *chromium trioxide* $(CrO_3)$ prepared in Exp. 7 in water; neutralise the acid solution by *potash*; evaporate, and allow to crystallise. Yellow crystals of potassium chromate, $K_2CrO_4$, separate out. Dissolve these crystals in water; divide the solution into two parts; to one part add $AgNO_3Aq$, and to the other $Pb2NO_3Aq$. In the one case a red pp. of silver chromate, $Ag_2CrO_4$, and in the other a yellow pp. of lead chromate, $PbCrO_4$, is obtained.

To obtain these salts of chromic acid ($H_2CrO_4$) from the basic oxide $Cr_2O_3$ it is only necessary to oxidise this compound in presence of a strongly basic oxide or hydroxide.

**Exp. 11.** Melt some *caustic potash* in a crucible over a Bunsen-lamp ; add a few crystals of *potassium nitrate* ($KNO_3$), and then a little $Cr_2O_3$. The $KNO_3$ is decomposed, giving off oxygen which oxidises the $Cr_2O_3$ to $2CrO_3$ and this reacts with the molten KOH to form $K_2CrO_4$. On cooling a yellow mass of $K_2CrO_4$ (mixed with unchanged KOH) is obtained.

*The production of yellow potassium chromate, $K_2CrO_4$, in this way is an extremely good test for the detection of chromium in salts of this metal.*

Besides the *chromates* there is known another series of salts called *dichromates*.

**Exp. 12.** To a saturated solution of *potassium chromate* add dilute *sulphuric acid* so long as the colour of the liquid changes ; then evaporate and allow to crystallise. Red-yellow crystals of potassium dichromate, $K_2Cr_2O_7$, separate out. ($2K_2CrO_4Aq + H_2SO_4Aq = H_2O + K_2Cr_2O_7Aq + K_2SO_4Aq.$)

In order to convert a chromate or dichromate into a salt of chromium it is necessary to heat with excess of a concentrated acid, or to treat with a *reducing agent*[*].

**Exp. 13.** To a concentrated solution of *potassium chromate* or *dichromate*, add a considerable excess of concentrated *sulphuric acid*, and evaporate the solution until the colour becomes deep green. Now prove that this solution gives the ordinary reactions of a chromic salt ; (1) *potash forms a greenish pp. of $Cr_2O_3.3H_2O$, soluble in considerable excess of the precipitant,* [compare this reaction with that between KOHAq and $CrO_3Aq$ (*Exp.* 10)]; (2) *ammonia precipitates the same $Cr_2O_3.3H_2O$, it is only slightly soluble in excess of ammonia.* The changes carried out in this Exp. may be thus represented in equations ;

$2K_2CrO_4Aq + 5H_2SO_4 = 2K_2SO_4Aq + Cr_2(SO_4)_3Aq + 5H_2O + 3O :$

$K_2Cr_2O_7Aq + 4H_2SO_4 = K_2SO_4Aq + Cr_2(SO_4)_3Aq + 4H_2O + 3O.$

A mixture of $K_2CrO_4$ or $K_2Cr_2O_7$ and $H_2SO_4$ will evidently act as an *oxidising agent*.[*]

[*] *s. Chap. XVII.*

**Exp. 14.**  To a solution of *potassium chromate* or *dichromate* add a little sulphuric acid, and then pass in *sulphur dioxide* (made by heating concentrated sulphuric acid with copper) until the liquid is green ; and prove that the solution now gives the reactions of a chromic salt.  The change which has occurred may be represented thus ; $2K_2CrO_4Aq + 2H_2SO_4Aq + 3SO_2 = 2K_2SO_4Aq + Cr_2(SO_4)_3Aq + 2H_2O.$

Two series of salts, *manganates* e.g. $K_2MnO_4$, and *permanganates* e.g. $KMnO_4$, may be obtained from the salts or the oxides of manganese.  As $MnO_2$ is practically insoluble in water, these salts are obtained by fusing this oxide with strongly basic oxides or hydroxides in presence of oxygen or an oxidising agent.  These salts do not correspond to the oxide $MnO_2$ but to a hypothetical oxide $MnO_3$.

**Exp. 15.**  Melt some *potash* in a crucible, add a very little *nitre*, and then a little *manganese dioxide*.  Keep the whole melted for a little and then allow to cool.  The green mass thus obtained contains potassium manganate, $K_2MnO_4$ : the reaction may be thus represented in an equation; $MnO_2 + 2KOH + O$ (from $KNO_3$) $= K_2MnO_4 + H_2O.$

*The production of green potassium manganate, $K_2MnO_4$, in this way is a delicate test for the presence of manganese in any compound.*

When the fused mass is cold, dissolve it in cold water, and add a few drops of dilute sulphuric acid ; the green colour changes to pink owing to the production of potassium permanganate.

Potassium permanganate interacts with sulphuric acid in the presence of an oxidisable body to produce potassium and manganese sulphates and oxygen ; the oxygen combines with the oxidisable body.

**Exp. 16.**  Make a solution of *ferrous sulphate* in cold water ; add some *sulphuric acid* and heat ; then add *potassium permanganate* solution ($KMnO_4Aq$) drop by drop with constant stirring until the liquid is coloured very faintly *pink*.  The pink colour is due to the presence of a minute trace of unchanged permanganate ; the permanence of this colour shews that the chemical change which has been occurring in the liquid is completed.  Now divide the liquid into two parts ; to one add a few drops of *potassium ferricyanide* solution ; the fact that no pp. is produced but only a slight brownish coloration shews that the liquid contains a ferric salt but is free from ferrous compounds.

But from the condition of the Exp. the only ferric salt that can be present is ferric sulphate. To the other part of the liquid add an excess of *ammonia*; the brown-red pp. which forms is a mixture of ferric and manganic hydrates; prove the presence of a manganese compound in this pp. by applying the fact that when these compounds are heated with molten potash and a very little nitre, green potassium manganate is produced (*s. Exp.* 15.) Exp. 2 taught us that manganic hydrate $Mn_2O_3.H_2O$ is pptd. when excess of ammonia is added to the solution of a manganese salt in presence of air and ammonium chloride; therefore the reaction with ammonia just concluded proves that the liquid obtained by adding the slightest possible excess of $KMnO_4Aq$ to $FeSO_4Aq$ in presence of $H_2SO_4Aq$ contained a salt of manganese; but from the condition of the Exp. this salt must almost certainly have been manganese sulphate $(MnSO_4)$.

We have therefore proved, as far as can be satisfactorily proved by qualitative experiments, that when potassium permanganate and ferrous sulphate interact in presence of sulphuric acid and water, ferric sulphate and manganese sulphate are produced. The quantitative study of the chemical change has shewn that it may be represented by the equation $10FeSO_4Aq + 8H_2SO_4Aq + 2KMnO_4Aq = 5Fe_23SO_4Aq + 2MnSO_4Aq + K_2SO_4Aq + 8H_2O.$

**Exp. 17.** To a solution of *oxalic acid* $(H_2C_2O_4)$ add *sulphuric acid*; warm, and add *potassium permanganate* solution, drop by drop, until a very slight pink colour remains in the liquid; now prove that the solution contains a manganese salt. To another quantity of warm oxalic acid solution mixed with sulphuric acid in a test tube arranged with a cork, exit tube, and funnel tube as shewn in fig. 35, add some $KMnO_4Aq$ and allow the escaping gas to pass into clear lime water; the production of a white pp. in the lime water proves that the gas evolved is carbon dioxide. The reaction which has occurred may be thus represented; $5H_2C_2O_4Aq + 3H_2SO_4Aq + 2KMnO_4Aq = K_2SO_4Aq + 2MnSO_4Aq + 8H_2O + 10CO_2.$

In Exps. 16 and 17, a salt of a manganese-containing acid $(KMnO_4)$ has been changed to a salt of manganese $(MnSO_4)$; the reverse change was accomplished in Exp. 15, taken in conjunction with Exp. 6. Compare the corresponding changes in the case of chromium compounds carried out in Exps. 11, 13, and 14.

The experiments performed in this chapter shew that the elements chromium manganese and iron are properly placed in

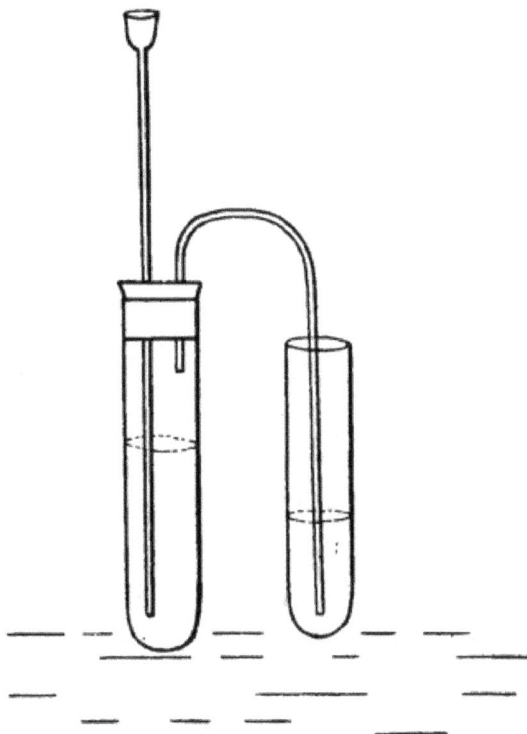

Fig. 35.

the same class; and that chromium and manganese are distinctly more negative and less metallic in their chemical reactions than iron.

*Reference to* "ELEMENTARY CHEMISTRY." Chap. XI. pars. 194—203.

# CHAPTER XVI.

THE products of a chemical change are often different at different temperatures.

**Exp. 1.** Dissolve two equal quantities of powdered *ferrous sulphate* in equal quantities of *water*, in one case using cold, and in the other hot, water. When solution is complete prove that the solution in cold water gives the reactions of a ferrous compound, but that the solution in hot water gives the reactions both of a ferrous and a ferric compound; part of the ferrous sulphate has been changed to (a basic) ferric sulphate.

**Exp. 2.** To two quantities of a solution of *copper sulphate*, one at the ordinary temperature, the other kept boiling, add a slight excess of *caustic potash* solution. In the cold solution a pp. of hydrated copper oxide ($CuO . H_2O$) is produced, but a pp. of copper oxide ($CuO$) is formed in the hot solution.

**Exp. 3.** To a cold solution of *formic acid* add a little *mercuric oxide* (prepared by precipitation); the oxide dissolves with formation of mercuric formate. To a hot solution of formic acid add mercuric oxide; a black pp. of mercury is produced.

In *Chap. XII. Exps.* 19 and 22, you learned that when chlorine and aqueous potash interact at ordinary temperatures potassium chloride and hypochlorite are produced, but that potassium chloride and chlorate are formed when the same bodies—chlorine and potash—interact at about 100°.

Sometimes the rate at which a chemical change proceeds

is dependent on the temperature, although the actual products of the change are the same at high as at low temperatures.

**Exp. 4.** Dissolve about 5 grams of *oxalic acid* in 300 to 400 c.c. of water, and add about 10—15 c.c. of concentrated *sulphuric acid.* Divide this solution into three equal parts; place these in beakers; keep one at the ordinary temperature, heat another to about 30°, and another to about 60°—70°. To each solution add the same quantity, about 5 c.c., of the same dilute aqueous solution of *potassium permanganate.* Note that the colour of the permanganate run into the hottest solution is at once destroyed, the colour of the permanganate run into the solution at 30° is destroyed after a very short time, and that the colour remains for some time in the cold solution. Repeat the addition of equal quantities of the same permanganate solution to these solutions, and note, approximately, the time which elapses in each case before the permanganate is completely decolorised. In each case the oxalic acid is oxidised to water and carbon dioxide, and manganese sulphate and potassium sulphate are produced (*s. Chap. XV. Exp.* 17).

**Exp. 5.** To each of three approximately equal quantities of a solution of *ammonium carbamate* ($CO.NH_2.ONH_4$) in cold water add a little *calcium chloride* solution; keep one solution at the ordinary temperature, warm another to about 30°, and another to about 100°. A pp. of calcium carbonate instantly forms in the hottest solution, after a little time in the solution at 30°, and only after a considerable time in the solution at the ordinary temperature. Ammonium carbamate interacts with water to form ammonium carbonate,

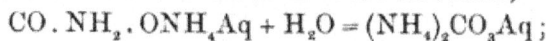

$$CO.NH_2.ONH_4Aq + H_2O = (NH_4)_2CO_3Aq;$$

this change takes place slowly at ordinary temperatures, and more rapidly the higher the temperature. In this Exp. the ammonium carbonate produced reacts with the $CaCl_2Aq$ to form $CaCO_3$ and $NH_4ClAq$, the former of which is precipitated.

Many chemical changes are completed only when the mass of one of the interacting bodies is large compared with the masses of the other interacting bodies. On the other hand when one of the products of an interaction between two bodies is completely insoluble, or volatile, under the experimental conditions, that product is usually produced, and the change is completed without using an *excess* of either of the reacting bodies.

**Exp. 6.** Place about 50 c.c. of dilute *sulphuric acid* in a beaker, add some *hydrochloric acid**, heat to boiling, and then slowly run in a solution of *barium chloride* from a burette so long as, but no longer than, a pp. of barium sulphate is produced. In order to determine when the reaction is completed, shake the contents of the beaker vigorously after the addition of each few c.c. of the $BaCl_2Aq$, allow the pp. to settle, and then add 1 or 2 drops of the $BaCl_2Aq$; when only a very slight pp. is produced, add 1 or 2 drops more $BaCl_2Aq$, shake well, allow to settle, pour a very little of the liquid through a filter (made of good paper), and add one drop of the $BaCl_2Aq$ to the clear filtrate; should the filtrate remain quite clear the reaction is completed, should a cloudiness be produced, run a drop or two of the $BaCl_2Aq$ into the beaker, and repeat the filtration &c. of a very little of the liquid; continue thus until the reaction is as nearly as possible completed, but do not add any *excess* of $BaCl_2Aq$.

When the reaction is complete, prove that the filtrate from the pp. of barium sulphate does not give the reactions for a compound of barium, or for sulphuric acid; prove also that the pp. is a barium compound and that it is a sulphate.

The reaction which has occurred is represented in an equation thus;—

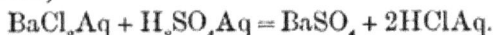

$$BaCl_2Aq + H_2SO_4Aq = BaSO_4 + 2HClAq.$$

The foregoing Exp. proves that when a given quantity of barium chloride in solution has been wholly decomposed by dilute sulphuric acid the only compound of barium produced is the sulphate which is precipitated, and that the sulphuric acid employed has been itself wholly decomposed. By making the Exp. more accurate it is possible to prove the correctness of the equation given above.

**Exp. 7.** You are given aqueous solutions of *barium chloride* and *sulphuric acid* of *stated strengths*, i.e. the mass of each compound in a specified volume of the solution is stated. Measure off accurately a certain volume of the barium chloride solution, say 50 c.c.; calculate exactly what volume of the acid solution contains that mass of sulphuric acid which is shewn by the equation in Exp. 6 to be necessary for the completion of the reaction; add exactly this volume of the sulphuric acid solution; allow the pp. to settle completely; pour

---

* The pp. of $BaSO_4$ settles better in the presence of HClAq.

the clear liquid through a filter, and prove the absence of a barium compound and of sulphuric acid, and the presence of hydrochloric acid, in the filtrate.

In the following Exps. solid compounds are produced, but they interact with the other products of the change to produce the original substances; hence it is necessary to add an excess of one of the interacting compounds in order to completely decompose the whole of the other compound in the original solution. Sometimes (*as in Exp.* 9) a very large excess of one of the interacting compounds is required to complete the change, even when that change results in the formation of a solid body.

**Exp. 8.** To a solution of *calcium chloride* add a solution of *ammonium carbonate* until the reaction is completed, i.e. until addition of a drop of ammonium carbonate solution fails to produce any pp. of calcium carbonate. Now filter, and prove that the filtrate contains a considerable quantity of ammonium carbonate. Hence some of the ammonium carbonate required to complete the chemical change has itself remained unchanged. The reaction may be represented by such an equation as the following:

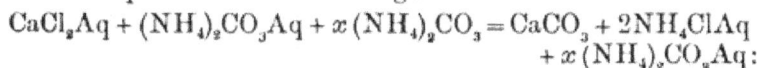

$$CaCl_2Aq + (NH_4)_2CO_3Aq + x(NH_4)_2CO_3 = CaCO_3 + 2NH_4ClAq$$
$$+ x(NH_4)_2CO_3Aq:$$

$x$ must be made fairly large to complete the reaction.

**Exp. 9.** To six equal quantities of a solution of *bismuth chloride* in dilute hydrochloric acid add quantities of *water*, in the ratio $1 : 2 : 4 : 6 : 8 : 10$*. In each case bismuth oxychloride is precipitated. Filter off each pp. and prove (by adding $H_2SAq$) that each filtrate contains bismuth chloride in solution. The reaction may be expressed thus:

$$BiCl_3 + x BiCl_3 + H_2O + Aq = BiOCl + 2HClAq + x BiCl_3Aq;$$

the smaller the value of $x$ relatively to that of Aq the greater is the quantity of $BiCl_3$ decomposed; but traces of $BiCl_3$ remain unchanged even when $x$ is several hundred times less than Aq.

**Exp. 10.** You are given solutions of *ferric chloride* and *potassium sulphocyanide*. To a few drops of one add a little

---

* About 10 c.c. of a solution of 10 grams $BiCl_3$ in 500 c.c. HClAq may be used: the smallest quantity of water may be about 10 c.c.

of the other; a deep red colour is produced, due to formation of ferric sulphocyanide. The reaction may be thus expressed:
$Fe_2Cl_6Aq + 6KCNSAq + xKCNSAq = Fe_2(CNS)_6Aq + 6KClAq + xKCNSAq$. If $x$ is made very large the reaction approaches completion.

Measure out 6 equal quantities—say 10 c.c. each—of the ferric chloride solution; add to the first a measured quantity $x$ c.c. (say 10 c.c.) of the sulphocyanide solution, to the second add $2x$ c.c., to the third $3x$ c.c., to the fourth $6x$ c.c., to the fifth $8x$ c.c., and to the sixth add $10x$ c.c., of the sulphocyanide solution. The depth of colour increases from the first to the sixth; this indicates that the quantity of ferric sulphocyanide produced has increased as the relative mass of potassium sulphocyanide was increased.

In the following Exp. a chemical change is more or less completely reversed by altering the relative masses of the interacting bodies.

**Exp. 11.** To some solid natural *antimony sulphide* (*Stibnite*) add a fair quantity of concentrated *hydrochloric acid*; antimony chloride is formed and passes into solution, while sulphuretted hydrogen gas is evolved. *Prove the latter fact by holding a piece of paper moistened with a solution of a salt of lead in the escaping gas; brown lead sulphide* (PbS) *is formed on the paper.* Now pour a considerable quantity of water into the vessel in which the reaction has occurred; orange-red antimony sulphide is precipitated. The two changes may be thus represented:—

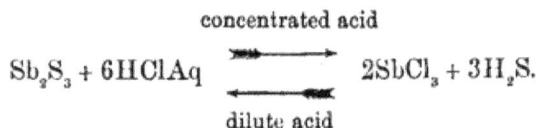

concentrated acid

$$Sb_2S_3 + 6HClAq \quad \rightleftharpoons \quad 2SbCl_3 + 3H_2S.$$

dilute acid

The $\rightarrow$ indicates the direction in which the change proceeds, according as much or little water is present.

In the following Exp. a chemical change is reversed by altering the medium in which the reaction occurs.

**Exp. 12.** Pass *carbon dioxide* into a saturated *alcoholic solution of dry potassium acetate*; potassium carbonate is formed and precipitated as a white solid, and acetic acid remains in solution.

To an *aqueous solution of potassium carbonate* add *acetic acid*; potassium acetate is formed and passes into solution, and carbon dioxide is evolved as a gas.

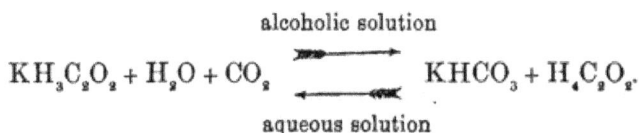

$$\text{KH}_3\text{C}_2\text{O}_2 + \text{H}_2\text{O} + \text{CO}_2 \underset{\text{aqueous solution}}{\overset{\text{alcoholic solution}}{\rightleftharpoons}} \text{KHCO}_3 + \text{H}_4\text{C}_2\text{O}_2.$$

**Exp. 13.** Recall *Exp. 13 Chap. XIV.* wherein nitric acid was prepared by the interaction of sulphuric acid with potassium nitrate. Place a small crystal of *potassium nitrate* in a porcelain basin, and a small crystal of *potassium sulphate* in another similar basin; to the first basin add some concentrated *sulphuric acid*, and to the second some concentrated *nitric acid*. Evaporate each to dryness, allow to cool, add more acid to each and again evaporate to dryness; repeat this treatment; and then prove that the salt in the first basin is potassium sulphate free from nitrate, and the salt in the second basin is potassium nitrate free from sulphate. The acid added in large excess has completely replaced the other acid from combination with the potash.

*Reference to* "ELEMENTARY CHEMISTRY." Chap. XII.

# CHAPTER XVII.

ANY chemical change which results in an increase in the ratio of the negative to the positive constituents of a compound is called an *oxidation*. Thus the change of FeO to $Fe_2O_3$, or of $FeSO_4$ to $Fe_2(SO_4)_3$, or of HgI to $HgI_2$, is a process of oxidation. Any change which results in an increase of the ratio of the positive to the negative constituents of a compound is called a *reduction*. Processes of oxidation and reduction usually occur together; one part of the complete change is called an oxidation, and the other part is called a reduction.

Oxidation is sometimes effected by the direct addition of oxygen.

**Exp. 1.** Cut a piece of *sodium* from the inside of a stick of this metal, and notice that the bright surface is at once covered with a non-lustrous film; this film is sodium oxide ($Na_2O$).

**Exp. 2.** Place a little very finely divided *iron* in a crucible, counterpoise the whole, then allow it to stand in the air for about an hour, and counterpoise again; there is no change of mass, nor is there any change in the appearance of the iron. Now heat the crucible over a Bunsen-lamp; the iron glows; allow to cool, and counterpoise again. There has been an increase in weight; the iron has been oxidised (to $Fe_2O_3$) by the oxygen of the air.

**Exp. 3.** Place a very small piece of *lead*, and a very small piece of *tin*, on charcoal, and direct the outer flame, i.e. the oxidising flame, of a blowpipe on to each in succession; the lead is slowly oxidised to a yellowish film of lead oxide, and the tin to a nearly white film of tin oxide.

But oxidations are more frequently accomplished by decomposing some compound of oxygen in contact with the body to be oxidised.

**Exp. 4.** Heat a few pieces of *charcoal* on an iron tray placed over a large Bunsen-lamp; then allow concentrated *nitric acid* to drop on to the hot charcoal. The charcoal is burnt (to carbon dioxide) by the oxygen furnished by the decomposition of the nitric acid (*s. Exp.* 13 of *Chap. XIV.*).

Recall *Exps.* 11 and 15 of *Chap. XIV.* wherein $Sb_2O_3$ was oxidised to $Sb_2O_4$ and $Sb_2O_5$, and phosphorus was oxidised to phosphoric acid, by interacting with concentrated nitric acid.

In all these experiments the nitric acid is reduced at the same time as the other body is oxidised.

**Exp. 5.** Heat a little *potassium chlorate* in a dry test tube until it melts, and apparently boils; then drop in a few small pieces of *charcoal*; notice the rapid burning of the charcoal, and prove that carbon dioxide is evolved. Allow to cool, dissolve the solid residue in water, and prove that the solution contains potassium chloride. The chlorate has been reduced, while the carbon has been simultaneously oxidised $[2KClO_3 \text{ (heated)} + 3C = 2KCl + 3CO_2]$.

**Exp. 6.** Heat a little *potassium nitrate* in a crucible until it melts, then throw in a few small pieces of *sulphur*; after cooling, dissolve in water and prove that the solution contains a sulphate.

The sulphur has been oxidised while the nitrate has been simultaneously reduced to nitrite $[KNO_3 \text{ (heated)} = KNO_2 + O]$.

Recall *Exp.* 15 of *Chap. XV.*, wherein a salt of manganese was oxidised by fusion with nitre in presence of potash.

**Exp. 7.** To a solution of *potassium nitrite* ($KNO_2$), acidulated with a little *sulphuric acid*, add a solution of *potassium permanganate*, drop by drop, until the permanence of a slight pink colour shews that a trace of permanganate remains unchanged. Now prove (1) the presence of a nitrate, (2) the absence of a nitrite, in this solution (*s.* tests in *Exp.* 4 *Chap. VII.* and *Exp.* 9, *Chap. XIV.*).

The potassium nitrite ($KNO_2$) has been oxidised to nitrate ($KNO_3$), and the permanganate ($KMnO_4$) has been reduced to MnO and $K_2O$ which have reacted with the sulphuric acid to form $K_2SO_4$ and $MnSO_4$ (*compare Exps.* 16 *and* 17 *of Chap. XV.*).

**Exp. 8.** Prepare a little very finely divided *sulphur* by adding hydrochloric acid to ammonium sulphide solution; now add some more HClAq, heat nearly to boiling, drop in a crystal of *potassium chlorate*, continue to heat and to add KClO₃ in very small successive quantities, until the sulphur has completely disappeared. Now prove that the solution contains a sulphate.

By the interaction of HClAq and KClO₃ a mixture of chlorine and oxides of chlorine is formed ; but chlorine oxides are very easily decomposed by heat to chlorine and oxygen ; hence in the foregoing process oxygen was freely evolved in contact with finely divided sulphur; the sulphur was thus oxidised to sulphuric acid. At the same time the potassium chlorate was reduced to chloride.

Recall *Exp.* 12 of *Chap. XIV.* wherein $Bi_2O_3$ was oxidised to $Bi_2O_5$ by forming and decomposing KClOAq in contact with the $Bi_2O_3$ in presence of much potash.

**Exp. 9.** To a solution of *potassium chlorate* add a few pieces of *zinc* and a little dilute *sulphuric acid*; hydrogen is evolved ; after a little filter some of the liquid and prove the presence of a chloride in the filtrate. The chlorate ($KClO_3$) has been reduced to chloride (KCl) by interacting with the hydrogen evolved in contact with the chlorate, and simultaneously the hydrogen has been oxidised to water.

**Exp. 10.** To a solution of *mercuric chloride* ($HgCl_2$) add a little *stannous chloride* solution ($SnCl_2$); a white pp. of mercurous chloride (HgCl) is formed ; now add more stannous chloride and heat; the colour of the pp. is changed from white to grey. This grey solid is very finely divided mercury ; by collecting it on a filter, drying, and rubbing in a mortar, globules of mercury are obtained. The reactions may be thus represented ;—

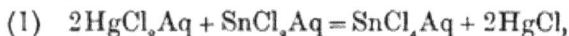

(1)    $2HgCl_2Aq + SnCl_2Aq = SnCl_4Aq + 2HgCl,$

(2)    $2HgCl + SnCl_2Aq = SnCl_4Aq + 2Hg.$

While the $HgCl_2$ is reduced to HgCl, and this is further reduced to Hg, the $SnCl_2$ is oxidised to $SnCl_4$.

**Exp. 11.** Add some *sulphuric acid* to a solution of *sodium sulphite* ($Na_2SO_3$) and prove that sulphur dioxide ($SO_2$) is evolved, by leading the gas into a dilute solution of potassium dichromate and noticing the green colour (due to the formation of chromic sulphate) which the liquid assumes (*comp. Exp.* 14, *Chap. XV.*).

Now place a few crystals of *sodium sulphite* in a little *ferric sulphate* solution, and add some *hydrochloric acid* ; heat to boiling, and continue to boil until only a very little $SO_2$ is coming off ($SO_2$ is easily detected by its smell). Then cool, and prove that the liquid contains ferrous sulphate.

In this reaction ferric sulphate ($Fe_2(SO_4)_3$) has been reduced to ferrous sulphate ($FeSO_4$), and sulphur dioxide ($SO_2$) has been simultaneously oxidised to sulphur trioxide ($SO_3$) which has interacted with the water present to produce sulphuric acid :—

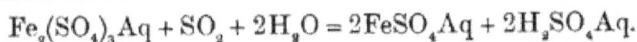

$$Fe_2(SO_4)_3Aq + SO_2 + 2H_2O = 2FeSO_4Aq + 2H_2SO_4Aq.$$

**Exp. 12.** Place a little *mercury* in a solution of *silver nitrate*; silver is slowly precipitated and some of the mercury passes into solution.

Place a piece of *copper* in a solution of *mercurous nitrate*; mercury is slowly precipitated and some of the copper passes into solution.

Place a piece of *iron* in a solution of *copper nitrate*, or *sulphate*; copper is slowly precipitated and some of the iron passes into solution.

These changes may be thus represented :—

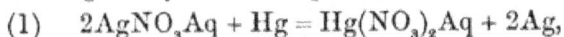

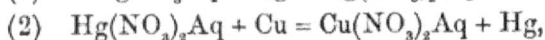

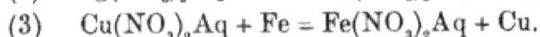

(1)   $2AgNO_3Aq + Hg = Hg(NO_3)_2Aq + 2Ag,$

(2)   $Hg(NO_3)_2Aq + Cu = Cu(NO_3)_2Aq + Hg,$

(3)   $Cu(NO_3)_2Aq + Fe = Fe(NO_3)_2Aq + Cu.$

In each case the metal placed in the solution reduces the salt in solution, the metal itself being simultaneously oxidised.

Compounds which readily part with oxygen, or with a part or the whole of their negative constituents, are often called *oxidising agents*; compounds, or elements, which readily remove oxygen, or negative elements such as chlorine &c., or negative groups of elements such as $SO_4$ &c., and combine with the oxygen &c. so removed, are usually called *reducing agents*.

*Reference to* " ELEMENTARY CHEMISTRY." Chap. XI. pars. 183 to 186.

# CHAPTER XVIII.

VARIOUS meanings have been given to the terms *strong* and *weak* as applied to acids and bases. The sense in which the terms are now generally used is shewn by considering the ordinarily occurring interaction between equivalent quantities of an alkali MOH and two acids HX and HY in dilute aqueous solution, to form two salts MX and MY in solution. In most cases portions of the alkali interact with each acid, so that when the interacting bodies have settled down into equilibrium the solution contains the two salts MX and MY, and also the two acids HX and HY. If the acid HX is a stronger acid than HY a greater quantity of the salt MX than of the salt MY will be formed when the change is completed. If for instance equivalent quantities of potash (KOH), nitric acid ($HNO_3$), and sulphuric acid ($H_2SO_4$), are mixed in dilute aqueous solution, about $\frac{2}{3}$ of the potash interacts with the nitric acid to form the salt $KNO_3$, and about $\frac{1}{3}$ of the potash interacts with the sulphuric acid to form the salt $K_2SO_4$; nitric acid is therefore said to be a stronger acid than sulphuric, and the relative strengths, or *affinities*, of the two acids for potash are said to be approximately in the ratio 2 : 1.

At a later stage of the course quantitative Exps. will be performed in illustration of the meaning of the terms strong and weak acids (*Part III. Chap. III.*); meanwhile Exps. will be conducted to shew that one acid can sometimes partially or wholly replace another from combination with a base, and to illustrate the conditions under which such replacement is effected.

**Exp. 1.** To two quantities of water coloured light yellow by addition of '*methyl orange*', add a little *boric acid*

($H_2B_2O_4$), and a little *sulphuric acid*, respectively. The boric acid does not change the colour of the liquid, but the sulphuric acid produces a bright red colour.

Now colour a concentrated solution of *borax* (sodium borate) with a little '*methyl orange*', and add *very dilute sulphuric acid* drop by drop until the colour changes to bright red. The fact that a considerable quantity of sulphuric acid must be added before the change of colour is produced shews that the sulphuric acid at first added was interacting with the borax to produce compounds which have no effect on the colour of '*methyl orange*'. But from what has been already learnt concerning the reactions between salts and acids, these products, very probably, were boric acid and sodium sulphate.

To a solution of *sodium sulphate* coloured by '*methyl orange*' add *boric acid*; no change of colour occurs even when a large quantity of boric acid has been added; hence boric acid does not interact with sodium sulphate, in aqueous solutions, to produce free sulphuric acid.

Sulphuric acid is therefore a stronger acid than boric.

**Exp. 2.** Dissolve a little *cane sugar* in water; add a few drops of *copper sulphate* solution, then a large excess of *potash*, and boil for a short time; no visible change occurs.

Repeat this Exp. with a *very little dextrose* (grape sugar) in place of cane sugar; a yellowish red pp. of cuprous oxide, $Cu_2O$, is at once produced.

To two approximately equal quantities of the same solution of *cane sugar*, in test tubes, add a few drops of dilute *hydrochloric acid* and of dilute *acetic acid*, respectively; immerse both tubes in water at 65° for 3 minutes; then add a few drops of *copper sulphate* solution and a large excess of *potash* to each, and boil. The formation of a reddish pp. in the tube to which hydrochloric acid was added, shews that this tube contains dextrose; the contents of the tube to which acetic acid was added remain unchanged.

Hydrochloric acid therefore changes cane sugar to dextrose when heated with it in dilute aqueous solution to about 65° for a few minutes; but acetic acid does not effect this change under similar conditions.

Now add a *few* drops of dilute *hydrochloric acid*, and a very little *cane sugar*, to a large quantity of a concentrated solution of *sodium acetate* ($NaH_3C_2O_2$); heat to 65° for 3 minutes, and then test for dextrose. The absence of dextrose proves that the whole of the hydrochloric acid added must have interacted with the sodium acetate so as to produce a compound of hydrochloric acid ($NaCl$ and $H_4C_2O_2$ are produced).

Hydrochloric acid is therefore a stronger acid than acetic.

**Exp. 3.**  Add *ferric chloride* solution to a solution of *sodium phosphate*; a yellowish white pp. of ferric phosphate ($FePO_4$) is formed.  Prove that this pp. is insoluble in acetic acid, but is easily dissolved by a small quantity of hydrochloric acid.

Now add a *few drops* of *hydrochloric acid* to a large quantity of a concentrated solution of *sodium acetate*; if the whole of the hydrochloric acid has interacted with the sodium acetate to produce sodium chloride and acetic acid, addition of sodium phosphate and ferric chloride should now produce a pp. of ferric phosphate; but if some of the hydrochloric acid remains in the free state no pp. of ferric phosphate should be formed. To the solution add a little *sodium phosphate* and a few drops of *ferric chloride* solution; ferric phosphate is precipitated, therefore the liquid contained no hydrochloric acid; therefore the hydrochloric acid added has combined with sodium previously combined with acetic acid, and therefore hydrochloric acid is a stronger acid than acetic.

One acid may wholly or partially replace another from its combinations with metals when the various bodies interact in aqueous solutions, and yet the second acid, if the less volatile of the two, may replace the first when the various bodies interact in the solid form at high temperatures.  The direction of the change which occurs at high temperatures between a salt and a solid acid depends to a great extent on the relative volatilities of the acids: the more volatile is displaced by the less volatile acid.

**Exp. 4.**  Pass a stream of *carbon dioxide* through a concentrated aqueous solution of *sodium silicate* (soluble glass); gelatinous silicic acid separates after a time, chiefly at the surface of the liquid.  Carbonic acid, then, displaces silicic acid from sodium silicate under ordinary conditions of temperature.

Prove that *ammonium chloride* solution ppts. silicic acid from an aqueous solution of *sodium silicate*.

Silicic acid is practically insoluble in water.

Mix a little *silica* (the insoluble anhydride of silicic acid) with 3 or 4 times its weight of a mixture of about equal parts *potassium carbonate* and *sodium carbonate* (this mixture melts at a lower temperature than either of its constituents) ; place the mixture in a crucible and heat until it is melted ; allow to cool ; dissolve in hot water ; filter if necessary from unchanged silica ; to the filtrate add *ammonium chloride* solution ; silicic acid is precipitated. Hence the solution contained sodium or potassium silicate, and therefore some of the carbonic acid has been displaced from combination with potash or soda by the silicic acid.

Silicic acid is non-volatile even at very high temperatures ; carbon dioxide is gaseous at ordinary temperatures.

**Exp. 5.** Prove (1) that a solution of *boric acid* slowly changes turmeric paper to a bright cherry-red ; (2) that *borax* solution (sodium borate) turns turmeric paper a dull brown ; (3) that when concentrated *hydrochloric acid* is added to *borax* solution and a piece of turmeric paper is steeped in this liquid for some time the colour of the paper is changed to cherry-red.

Mix a little solid *boric acid* with 3 or 4 times its weight of solid *potassium sulphate*, and strongly heat the mixture in a crucible, stirring the contents with a rod from time to time. Then dissolve in cold water and place a piece of *turmeric paper* in the solution ; the colour is not changed to red until concentrated hydrochloric acid is added ; hence the boric acid has decomposed the potassium sulphate to form potassium borate.

Compare this result with that of *Exp. 1 of this Chap.*

**Exp. 6.** Prepare a solution of *albumen* by shaking up a little white of egg with water and filtering.

Prove (1) that a solution of *metaphosphoric acid* ($HPO_3$) ppts. albumen from this liquid ;

(2) that *acetic acid* does not ppt. albumen ;

(3) that *sodium metaphosphate* ($NaPO_3$) solution does not ppt. albumen ;

(4) that a solution of *sodium metaphosphate* to which *acetic acid* has been added does ppt. albumen.

These reactions shew that acetic acid in aqueous solution is a stronger acid than metaphosphoric.

Now make a small spiral on the end of a piece of platinum-wire ; place a little solid *metaphosphoric acid* in this spiral, heat it, and dip the melted mass into powdered *potassium sulphate*; repeat this, until the bead has become nearly infusible in the Bunsen-flame. Dissolve the solid on the wire in water and add albumen solution; the albumen is not precipitated; now add some acetic acid, the albumen is precipitated.

These reactions shew that solid metaphosphoric acid displaces sulphuric acid from potassium sulphate when this salt is strongly heated with the acid. Sulphuric acid is volatile, metaphosphoric acid is non-volatile, at a red heat.

*Reference to* " ELEMENTARY CHEMISTRY." Chap. XIII.

The student should now be required to devise and conduct experiments bearing on the subjects treated in Chaps. I. to XVIII., and to reason on the results he obtains. That he may do this satisfactorily, it will sometimes be advisable to set before him the outlines of suitable experiments in the form of questions, which he must answer by experimentally applying the knowledge gained by his work in the laboratory and his reading in connexion with that work. Appendix I. contains a few questions which may be found useful. Appendix II. contains a few questions rather more difficult than those in App. I.; some acquaintance with volumetric analysis is required to work out the questions in App. II.

The student should now work through a course of qualitative analysis sufficient to enable him to analyse qualitatively easy mixtures of metallic salts. This course should be followed by one comprising the more important processes of quantitative volumetric (or titrimetric) analysis, and a few simple gravimetric analyses.

A few tables are given in Appendix III. which may be useful in qualitative analysis; but these tables must not be used until the student has conducted experiments illustrative of the principles on which the tables are framed, and has made himself practically acquainted with the reactions which are used as tests for the various metals and acids.

*The experiments described in Part II. of this book assume that the student has a fair practical acquaintance with qualitative, and quantitative volumetric, analysis; and that he can accurately perform the ordinary processes of gravimetric analysis, such as the manipulation of precipitates, weighing, &c.*

# PART II.

## CHAPTER I.

Law of multiple proportions. *When two elements combine to form more than one compound, the masses of one of the elements which combine with a constant mass of the other element bear a simple relation to each other.*

Determine the ratio of oxygen to nitrogen in (1) *nitrous oxide*, (2) *nitric oxide*, each of which is a compound of the two elements oxygen and nitrogen.

**Exp. 1.** Arrange an apparatus as shewn in Fig. 36. *A* is a small hard glass flask; this is connected with two bottles con-

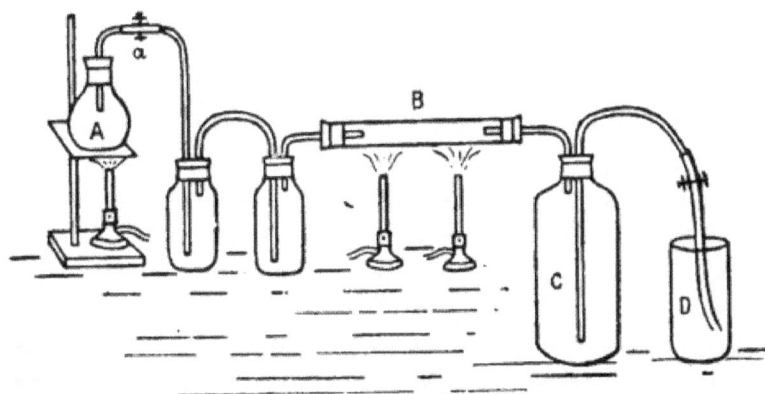

Fig. 36.

taining concentrated *sulphuric acid*; at *a* is a piece of caoutchouc tubing with a screw clamp; the exit tube from the second of these drying bottles passes into a piece of hard glass tubing about 15 centims. long (*B*) arranged so that it can be heated by a couple of Bunsen-lamps; the other end of this tube is connected with a large bottle of about 2 litres capacity (*C*); another tube passes from the bottom of this bottle through the cork and is connected with a piece of caoutchouc tubing furnished with a small screw-clamp. The cork of *C* must fit very tightly. *D* is a *perfectly dry* beaker capable of holding about 500 c.c.

Place about 15 grms. of *dry ammonium nitrate* in *A*; loosely fill *B* with well *cleaned* and *perfectly dry copper* turnings; completely fill *C* with water; suck at the end of the caoutchouc tube until a few drops of water flow out, and then firmly close the screw-clamp. Now remove the corks from *B*; place a cork in each end of the tube, and weigh it with its contents, and then replace it. Heat *B* gradually until it is at a full red heat; while this is being done disconnect *A* from the rest of the apparatus at the point *a*; cautiously heat the ammonium nitrate in *A* until it melts and evolves gas, but do not raise the temperature higher than suffices to cause the evolution of a fairly rapid stream of gas; after a few minutes connect *A* with the rest of the apparatus, and at once open the screw-clamp so as to allow water to drop quickly, but not to flow, from *C* into the beaker *D*. The tube *B* must be kept at a red heat while these things are being done. Nitrous oxide is produced in *A* ($NH_4NO_3$ heated $= N_2O + 2H_2O$; if the temperature is allowed to get too high other changes are produced). The nitrous oxide is dried by the concentrated sulphuric acid, and is then decomposed by the hot copper in *B*, with formation of nitrogen and copper oxide ($Cu + N_2O$ heated $= CuO + 2N$); the nitrogen collects in the bottle *C*, the copper oxide remains in *B*.

The progress of the decomposition in *B* may be traced by observing the change of colour from red copper to black copper oxide; care must be taken that the copper at the end of *B* nearest *C* remains bright and unoxidised at the close of the experiment, for should the whole of the copper be oxidised a portion of the nitrous oxide will almost certainly pass unchanged into *C*. Care must also be taken that the end of the caoutchouc tube in *D* remains beneath the surface of the water in *D*. When about 200—300 c.c. of nitrogen have collected in *C* the lamps are removed; when the stream of gas has slackened the screw-clamp at *a* is firmly closed and the cork is instantly removed from *A*; the apparatus is now allowed to cool; the corks are removed from $B$, and those which closed the ends of the tube when it was previously weighed are put in their places. The water in the beaker $D$ is carefully measured in c.c.; the temperature of the air, and the barometric pressure are noted; and $B$ and its contents are weighed.

The increase in the weight of $B$ gives the mass of oxygen obtained from the nitrous oxide which passed over the hot copper. The volume of water in $D$ is equal to the volume of nitrogen obtained from the nitrous oxide; this volume is reduced to $0^{\circ}$ and 760 mm., a correction being made for the pressure of aqueous vapour at the temperature of observation; and the weight in grams of nitrogen is found by multiplying the corrected volume in c.c. at $0^{\circ}$ and 760 mm. by ·001257, inasmuch as 1 c.c. of nitrogen at $0^{\circ}$ and 760 mm. weighs ·001257 grams.

From the results obtained calculate the mass of nitrogen which combines with 8 parts by weight of oxygen to form nitrous oxide.

**Exp. 2.** Modify the apparatus used in Exp. 1 by putting a larger flask, fitted with a safety funnel, in place of $A$. Arrange the rest of the apparatus as in Exp. 1; weigh $B$ with its contents; fill $C$ with water, &c. as in Exp. 1. Into the flask $A$ put about 10 grams of *copper turnings*, cover these with water, and add *nitric acid* (spec. grav. about 1·2) by the funnel tube until a fairly rapid stream of gas is evolved; allow the first portion of the gas to pass away at $a$. Nitric oxide is produced by the interaction of copper and nitric acid $3Cu + 8HNO_3Aq = 3Cu(NO_3)_2Aq + 4H_2O + 2NO$; when nitric oxide is mixed with oxygen a dark red gas—$NO_2$ and $N_2O_4$—is produced. As soon as the gas filling the flask $A$ is quite colourless and the copper in $B$ is red hot, connect the exit tube from $A$ with the rest of the apparatus, and open the screw-clamp so that water drops quickly, but does not flow, from $C$. When the greater part of the copper in $B$ is oxidised remove the lamps; close the screw-clamp at $a$, and instantly remove the cork from the flask $A$; allow everything to cool; weigh $B$ and its contents; and measure the water in $D$. Nitric oxide is completely decomposed by hot copper to nitrogen and oxygen; the oxygen combines with the copper to form copper oxide.

From the results obtained calculate the mass of nitrogen

which combines with 8 parts by weight of oxygen to form nitric oxide*.

If the experiments are carefully conducted, you find that the masses of nitrogen which combine with a fixed mass of oxygen to form nitrous and nitric oxides are in the ratio 2 : 1.

*Reference to* "ELEMENTARY CHEMISTRY." Chap V. pars 54 to 64.

Law of reciprocal proportions. *The masses of different elements which severally combine with one and the same mass of another element are also the masses of these elements which combine with each other, or they bear a simple relation to those masses.*

Prepare and analyse a compound of each of the elements sulphur and chlorine with the element lead, and then a compound of sulphur and chlorine with each other.

**Exp. 3.**    Prepare and analyse *lead sulphide.*
Dissolve about 30 grams of pure *lead nitrate* in cold water; add one or two drops of dilute *nitric acid*; warm gently, and pass a slow stream of well washed *sulphuretted hydrogen* into the liquid until it smells strongly of this gas. Warm the liquid holding lead sulphide in suspension to about 60°; after 10 or 15 minutes again pass in sulphuretted hydrogen for some time. Now heat the contents of the beaker nearly to boiling; allow the pp. of lead sulphide to settle; wash it repeatedly, by decantation, with warm water; then transfer the pp. to a basin and dry it over sulphuric acid *in vacuo.*
Weigh out accurately, into a large weighed porcelain crucible, about ·5 gram of the lead sulphide you have prepared. Place the crucible on a water-bath and add a few drops of very concentrated fuming nitric acid; heat on the bath until all visible action ceases; then add a very little more nitric acid, continuing to heat on the bath. Continue this treatment, taking care to add the concentrated nitric acid very slowly, until not a trace of unchanged black lead sulphide is visible in the crucible; add no more acid, but heat on the bath till the contents of the crucible are dry; then heat over a small Bunsen-flame gradually raising the temperature till the bottom of the crucible is red hot, taking care that the gases

---

* Exps. 1 and 2 are taken with slight modifications from Ramsay's *Experimental Proofs of Chemical Theory.*

produced in the flame of the Bunsen-lamp do not come into contact with the contents of the crucible. After a few minutes allow to cool, and weigh.

The lead sulphide is oxidised, at the expense of the nitric acid, to lead sulphate.

From the weight of lead sulphate found, calculate the ratio of lead to sulphur in the lead sulphide, assuming that the substance you prepared was a compound of lead and sulphur, and of these elements only, and that the composition of lead sulphate is expressed by the formula $PbSO_4$.

Lead and sulphur combine to form lead sulphide in the ratio 1 : ·155.

**Exp. 4.**  Prepare and analyse *lead chloride*.

Dissolve about 20 grams of pure *lead nitrate* in as little cold water as possible; add pure dilute *hydrochloric acid* little by little as long as a white pp. of lead chloride is formed. Allow the pp. to settle; recrystallise repeatedly from hot water; then transfer the pp. to a basin and dry at 100°.

Weigh out accurately, into a large weighed porcelain crucible, about ·5 gram of the lead chloride you have prepared; and convert it into lead sulphate by treatment with *pure* concentrated sulphuric acid, added little by little to the lead chloride heated on a water-bath, in much the same way as directed in Exp. 3. Finally heat to redness and weigh the lead sulphate. The hydrochloric acid produced in the reaction ($PbCl_2 + H_2SO_4 = PbSO_4 + 2HCl$) is volatilised.

Assuming that the substance you prepared was a compound of lead and chlorine, and of these elements only, it is easy to calculate the ratio of chlorine to lead from the results of your analysis.

Lead and chlorine combine to form lead chloride in the ratio 1 : ·343.

**Exp. 5.**  Prepare and analyse *sulphur chloride*.

Place about 30 grams of *sulphur* in small pieces in a *dry* retort the beak of which passes into a small *dry* flask kept cold by a stream of running water. Fit the neck of the retort with a good cork through which passes a tube coming from an apparatus for evolving *perfectly dry chlorine*. Heat the sulphur, and pass dry chlorine into the retort; the chlorine is rapidly absorbed and a yellow liquid is formed. When the contents of the retort are quite liquid stop the stream of chlorine, put a little sulphur into the retort, and heat until the greater part of the liquid has distilled over into the receiver.

Pour the liquid from the receiver into a distilling flask; connect this flask with a condenser and receiver; fit a thermometer into the flask, and distil; have 3 or 4 small *dry* flasks ready, and collect the liquid which boils between 136°

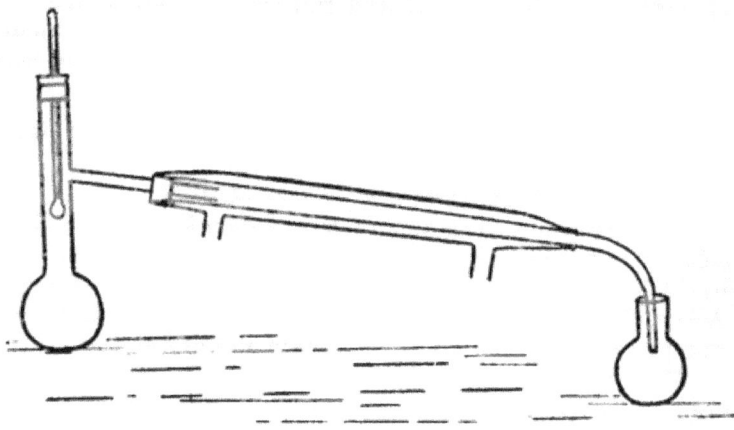

Fig. 37.

and 140° separately from the rest (*s.* Fig. 37). Redistil this liquid, collecting that which boils from 137° to 138°; this is nearly pure sulphur chloride.

The vapour of sulphur chloride affects the eyes and the mucous membranes; care must be taken in working with this compound.

Prepare a small bulb of thin glass with a drawn out neck about the size shewn in Fig. 38. Weigh this bulb accurately; warm the bulb for a few seconds, and at once plunge the open end into the sulphur chloride you have prepared; the liquid is forced up into the bulb and nearly fills it. Now close the open end of the bulb by holding it for an instant in the flame of a Bunsen-lamp; allow it to cool, and weigh; you thus find the weight of sulphur chloride in the bulb. Place the bulb in about 300 c.c. of water; break the bulb by the help of a glass rod; add a good deal of nitric acid,

Fig. 38.

warm, filter if necessary, and then add silver nitrate solution so long as a pp. of silver chloride is formed: wash this pp. thoroughly with hot water by decantation; then collect and weigh it. (For details a manual of quantitative analysis must be consulted.)

Sulphur chloride interacts with water to form hydrochloric acid, sulphur dioxide, and sulphur.

From the weight of silver chloride found, the weight of chlorine in the quantity of sulphur chloride taken is calculated, and from this the ratio of chlorine to sulphur is deduced, assuming the substance you have prepared to be a compound of sulphur and chlorine only.

Sulphur and chlorine combine to form sulphur chloride in the ratio 1 : 1·11.

State the results of Exps. 3, 4, and 5 so as to shew (1) the mass of sulphur you have found combined with 1 part by weight of lead, (2) the mass of chlorine you have found combined with 1 part by weight of lead, (3) the mass of chlorine you have found combined with that mass of sulphur which combined with 1 part by weight of lead. Then shew how your results illustrate the law of reciprocal proportions.

*Reference to* "ELEMENTARY CHEMISTRY." Chap. V. pars. 66 to 73.

# CHAPTER II.

## EQUIVALENT AND COMBINING WEIGHTS.

THE *equivalent weights* of a series of elements or compounds must be determined with relation to some one defined chemical change; those masses of the various bodies which severally bring about, or take part in, this specified change are said to be equivalent.

. The *combining weight* of an element is usually taken to be that mass of it which combines with, or replaces, 1 part by weight of hydrogen, 8 parts by weight of oxygen, 16 parts by weight of sulphur, or 35·5 parts by weight of chlorine.

**Exp. 1.** Determine the masses of *aluminium, magnesium, zinc,* and *tin,* which severally interact with *hydrochloric acid* to produce 1 gram of hydrogen.

Procure three glass tubes, each closed at one end, capable of holding 100 c.c., and graduated to $\frac{1}{10}$ c.c. Perfectly fill these tubes with dilute hydrochloric acid, about 1 part concentrated acid to 2 parts water, invert each in a small basin of dilute hydrochloric acid, and firmly clamp each tube. Now weigh out accurately small pieces of aluminium, magnesium, and zinc, in the form of thin foil; take about ·05 gram of aluminium, about ·06 gram of magnesium, and about ·2 gram of zinc\*. Put each little bit of metal into a very small glass tube closed at one end, about the size shewn in Fig. 39, and fill these tubes with water. Close each tube with a loosely fitting plug of glass-wool. Press the finger over the open end of each tube, taking care that no air bubbles are admitted;

Fig. 39.

\* These quantities will evolve something like 70 c.c. of hydrogen from hydrochloric acid.

bring each tube under the acid in the basin and beneath the opening of the graduated tube, then remove the finger, and lower the graduated tube so that the small tube containing the metal cannot escape. The acid and metals interact; the hydrogen evolved collects in the graduated tubes. When the reactions have wholly ceased, move each tube and basin into a wide deep vessel containing water; remove the basins and lower the tubes until the levels of the liquid inside and outside are the same; then read off the volume of hydrogen in each tube. Determine the temperature and pressure of the air; make the necessary correction for the pressure of water-vapour at the temperature of observation.

From the data thus obtained calculate the masses of the three metals which have severally produced 1 gram of hydrogen by reacting with hydrochloric acid. These masses of the three metals are equivalent inasmuch as they bring about equal amounts of a certain chemical change, viz. replacement of hydrogen from hydrochloric acid.

The numbers ought to be 9 for aluminium, 12 for magnesium, and 32·5 for zinc.

The foregoing method is not applicable to the case of *tin*, which is dissolved by *cold* hydrochloric acid only with extreme slowness.

Arrange an apparatus as shewn in fig. 40. The small flask *A* is fitted with a sound cork, the under surface of which must be perfectly smooth. Through this cork pass an entrance tube *a* connected by caoutchouc tubing with the burette *B*, and an exit tube *b*. The bore of the exit tube is contracted at *d* and *e*, so as to keep in position a plug of glass-wool placed between the contracted parts. The extremity *d* of the exit tube must be flush with the under-surface of the cork. To use the apparatus place an accurately weighed quantity of tin foil (about ·2 to ·3 gram) in the flask *A*, pour on it a few cubic centims. of a concentrated solution of *platinic chloride*\*, then fill the flask completely with air-free water and cork it. Fill *B* also with air-free water and turn on the tap *c* till the stream of water from the burette has completely expelled all the air from the flask and the exit tube. Now bring the collecting tube *C* into position on the bee-hive; empty *B* of water and

---

\* Tin precipitates metallic platinum from the chloride, and the platinum-tin couple thus obtained interacts more readily with the acid than tin alone.

refill it with concentrated hydrochloric acid.   From time to time allow the acid to flow into the flask and heat the contents of

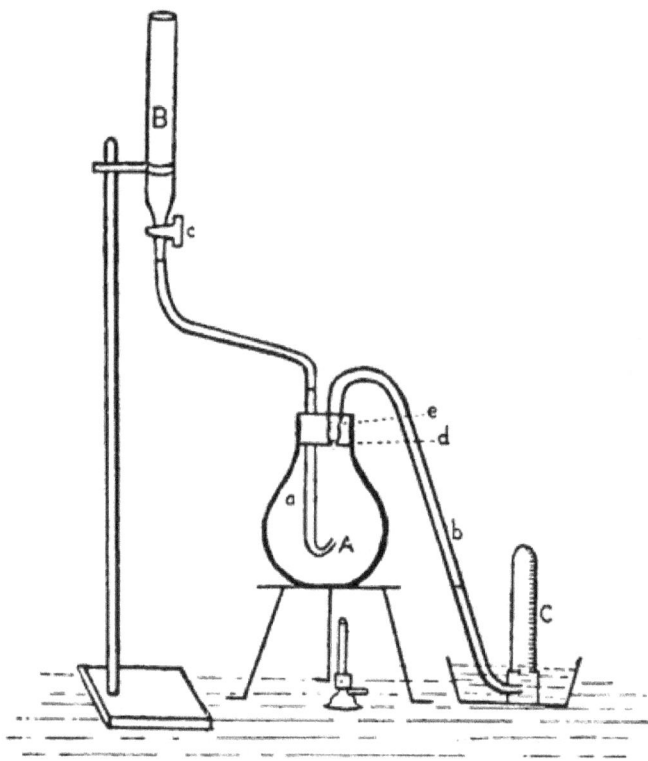

Fig. 40.

the latter to boiling.   When hydrogen is no longer evolved, again fill the apparatus completely with water from *B*. Reduce the volume of hydrogen in *C* to $0^{\circ}$ and 760 mm., and calculate the mass of tin which has produced 1 gram of hydrogen by reacting with hydrochloric acid.

The result should be 59 grams of tin.

**Exp. 2.**   Determine the masses of *magnesium* and *tin* which severally combine with 8 parts by weight of *oxygen*.

Accurately weigh a large porcelain crucible with its lid; weigh out about ·5 gram of thin *tin-foil* into the crucible; place the crucible on a sand-tray, and add *nitric acid*, a drop or two at a time, so long as any visible action occurs.   The tin is oxidised to stannic oxide, and water and oxides of nitrogen escape as gases. When the whole of the tin has disappeared and a white powdery mass has taken its place, put the lid on the crucible, and heat the sand-tray gradually to full redness, removing the lid for a second or two from time to time ; then remove the crucible to a triangle and heat strongly for a few minutes. Allow to cool and weigh.

From the results obtained, calculate the mass of tin which has combined with 8 parts by weight of oxygen.

Arrange an apparatus as shewn in Fig. 41. $A$ is a small dry bulb of hard glass connected with two small light U tubes $B$ and $C$ ; the limb of $B$ next the bulb $A$ contains clean broken glass thoroughly wetted with water; $a$ is a plug of loosely packed glass-wool ; the other limb of $B$ and the tube $C$

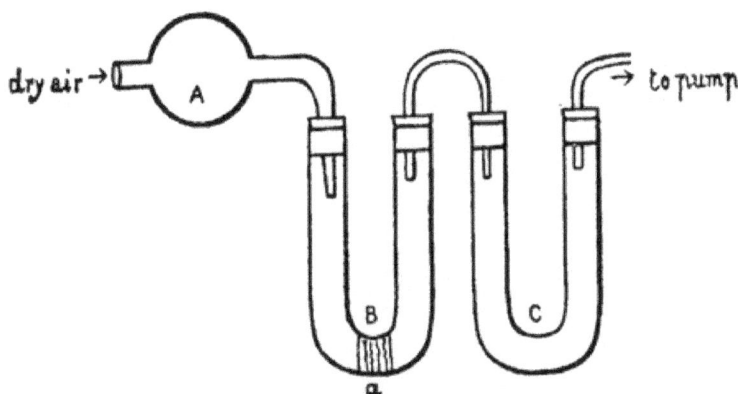

Fig. 41.

contain calcium chloride. The open end of the bulb $A$ communicates with two large U tubes containing calcium chloride, and these with a bottle containing potash solution. The exit tube from $C$ is in connection with a water-pump : when the pump is set in action a stream of air, freed from carbon dioxide and moisture, is sucked through the whole apparatus.

Weigh the whole apparatus. Then place about ·3 to ·4 gram (accurately weighed) of magnesium ribbon in $A$; heat gently while a slow air-current is kept passing through the apparatus, then raise the temperature until the magnesium burns. By regulating the rate of the air-current, and the temperature to which $A$ is raised, the magnesium may be completely burnt to oxide and the whole of the oxide retained in $A$ and in the tubes $B$ and $C$; the calcium chloride serves to retain any moisture which may be volatilised from the wet glass in $B$. When the combustion is finished allow the apparatus to cool, and then weigh it.

Let the weight of the apparatus before the experiment be $x$ grams, the weight of magnesium burnt be $y$ grams, and the weight of the apparatus and magnesia after the experiment be $z$ grams, then the sum of $x$ and $y$ deducted from $z$ gives the grams of oxygen which have combined with the $y$ grams of magnesium to form magnesia.

From your results calculate the mass of magnesium which combines with 8 parts by weight of oxygen to form magnesium oxide.

The results ought to shew that 29·5 parts by weight of tin, and 12 parts by weight of magnesium, severally combine with 8 parts by weight of oxygen.

A comparison of these results with those obtained in Exp. 1 shews (1) that the mass of magnesium which evolves 1 part by weight of hydrogen from hydrochloric acid is also the mass which combines with 8 parts by weight of oxygen; and (2) that the mass of tin which evolves 1 part by weight of hydrogen from hydrochloric acid is a simple multiple of the mass which combines with 8 parts by weight of oxygen to form stannic oxide.

Tin forms another compound with oxygen (stannous oxide) in which tin and oxygen are combined in the ratio 59 : 8.

**Exp. 3.**    Determine the combining weight of *silver* by finding the mass of this metal which combines with 8 parts by weight of *oxygen* to form silver oxide.

Prepare silver oxide; then decompose a weighed quantity of it by heat, weigh the silver which remains, and measure the volume of oxygen evolved.

Dissolve about 6 grams of pure *silver nitrate* in water in a flask; pour *freshly prepared clear lime water* into the flask, through a filter arranged so that the lower end of the filter dips under the surface of the nitrate of silver solution; continue to add lime water with constant shaking so long as any pp. forms; perform this process as quickly as possible, because if moist silver oxide is exposed to the air for any length of time it interacts with the carbon dioxide of the air to form silver carbonate. When precipitation is complete, close the flask with a cork, and shake the contents violently; allow the pp. to settle, pour off the clear liquid, and wash the pp. as quickly as possible with hot water, by decantation, until the washings nearly cease to shew an alkaline reaction with litmus and are quite free from nitrates.

The reaction occurring in the flask is thus represented in an equation $2AgNO_3Aq + CaOAq = Ag_2O + Ca(NO_3)_2Aq$.

When the silver oxide is thoroughly washed transfer it to a basin, and dry it at $100^{\circ}$.

Weigh out accurately from $2\frac{1}{2}$ to 3 grams of the dry silver oxide into a piece of hard glass tubing about 15 centims. long, closed at one end, which has been accurately weighed. Fit this tube with a very good tightly-fitting cork through which passes an exit tube as shewn in Fig. 42. Let the exit tube

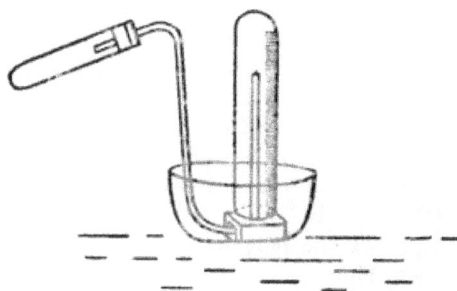

Fig. 42.

pass under the surface of water about half way up into a graduated 200 c.c. tube filled with water.

Gently heat the tube, beginning a little way above the upper surface of the silver oxide and gradually heating downwards to the bottom of the tube; slowly raise the temperature until the contents of the tube are at a full red heat. Continue to heat so long as any oxygen is evolved. The open end of the exit tube will now be above the surface of the water in the graduated tube. Remove the lamp and allow the apparatus to cool.

Weigh the hard glass tube with its contents; the difference between this weight and that of the empty tube gives the weight of silver obtained from the weight of silver oxide taken. Assuming that the compound you prepared was pure silver oxide, calculate the mass of silver combined with 8 parts by weight of oxygen.

Read off the volume of oxygen in the graduated tube after equalising the pressures inside and outside as already described in Exp. 1; make the necessary corrections for temperature, pressure, and pressure of water-vapour, and then find the weight of oxygen obtained. Now calculate the mass of silver combined with 8 parts by weight of oxygen in silver oxide.

If the experiments are carefully conducted you find that 108 parts by weight of silver combine with 8 parts by weight of oxygen to form silver oxide.

**Exp. 4.**  *Oxalic acid* is easily oxidised to carbon dioxide and water; *ferrous sulphate* is easily oxidised to ferric sulphate. Taking those masses of oxalic acid and ferrous sulphate which are oxidised by one and the same mass of oxygen to be equivalent one to the other, determine the equivalent of ferrous sulphate, that of oxalic acid being taken as unity.

Weigh out accurately into beakers about ·5 gram of pure crystallised *oxalic acid* and about ·5 gram of pure *ferrous sulphate*. Dissolve each in water, add a good deal of concentrated *sulphuric acid*, warm, and at once run in a standardised solution of *potassium permanganate* until each oxidation is completed. (For details *s.* a manual of quantitative analysis.)

As the strength of the permanganate solution is known, it is easy to calculate the masses of ferrous sulphate and oxalic acid oxidised by a specified mass of oxygen, and as these masses are equivalent as regards this reaction, the equivalent of one of the compounds is readily found when that of the other is taken as unity.

One part by weight of crystallised oxalic acid ($H_2C_2O_4 . 2H_2O$) is equivalent to 2·41 parts of ferrous sulphate ($FeSO_4$) as regards quantity of oxygen required to oxidise both compounds.

*Reference to* "ELEMENTARY CHEMISTRY." Chap. V. pars 73 to 75; also Chap. XVII. pars. 346 to 350.

# CHAPTER III.

[Before beginning the experiments in this chapter the student ought to read chapter XV. of " ELEMENTARY CHEMISTRY."]

To determine the *molecular weight* of an element or compound it is necessary to find the specific gravity of the body as a gas. The product obtained by multiplying the specific gravity of the gas, referred to air as unity, by 28·87 is the specific gravity referred to hydrogen as 2, in other words is (approximately) the molecular weight of the gas. The molecular weight of a compound gas must be equal to the sum, or to a multiple of the sum, of the combining weights of the elementary constituents of the compound ; hence the data required for the accurate determination of the molecular weight of a compound are, (1) accurate determinations of the combining weights of the elementary constituents, and (2) an approximately accurate determination of the specific gravity of the compound as a gas. (*s. especially pars.* 294 *to* 300, *Chap. XV. of* "ELEMENTARY CHEMISTRY.")

The method for determining the specific gravities of gasifiable bodies introduced by v. Meyer is generally used.

**Exp. 1.** Determine the *specific gravities of gaseous ether, alcohol, and water,* and from your results find approximate values for the molecular weights of these compounds.

As will be seen from the annexed figure (Fig. 43) the apparatus consists essentially of a large cylindrical bulb *b* of about 80 c.c. capacity, fused on to the lower end of a glass tube about 500 mm. long, which has fused to its upper end an enlarged mouth-piece, provided with a well-fitting cork

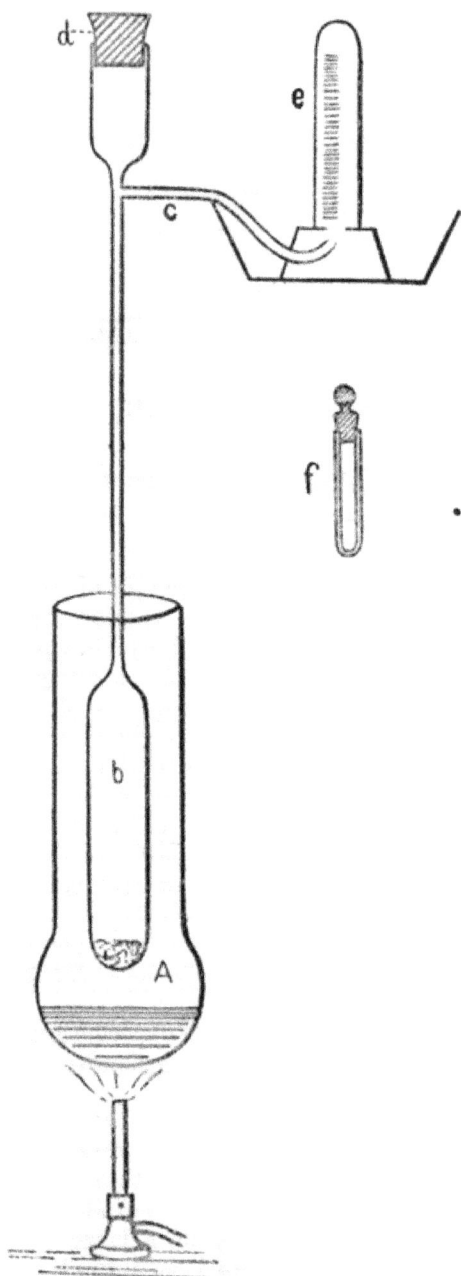

Fig. 43.

*d.* Just beneath the enlargement, a capillary side-tube *c* is given off. At the bottom of the bulb *b* is a small cushion of dry glass wool, or asbestos. The apparatus must be thoroughly dried before use, by blowing air into it with a bellows while it is gently heated.

(1) To determine the vapour density of ether. Water is caused to boil briskly in the cylindrical glass bath *A*, and the dried apparatus, with its mouth corked, is so clamped that the bulb *b* is well surrounded by the steam in *A*, while the capillary *c* dips into a small pneumatic trough. The air in *b* expands and escapes through *c*; after a time equilibrium is attained, and no more air escapes. When this is the case, the graduated cylinder *e* is brought over the bee-hive; the cork *d* is removed, the little flask *f* containing the ether is dropped into the apparatus, and the cork *d* is instantly replaced. The dry flask *f* has been weighed, and then again weighed when *completely* filled with ether; the flask should contain about ·05 gram of ether. The difference in the two weighings (*w* grms.) gives the weight of ether used. The ether is quickly and completely vaporised in *b*, and the vapour displaces a corresponding volume of air through *c*; this air is collected in the graduated tube *e*. When the vaporisation is complete, as indicated by the cessation of air bubbles from *c*, the tube *e* is removed by means of a spoon to a vessel containing water, in which it is lowered until the levels of the water inside and outside the tube are coincident. The volume in c.c. *V*, the temperature *t*, and the pressure *p*, are then read; and the pressure of aqueous vapour *f* at the temperature *t* is noted.

If the measurement had been made at 0° and 760 mm. the volume of air would have been

$$\frac{V}{1 + at} \times \frac{p - f}{760} \text{ c.c.};$$

where $a$ = coefficient of expansion of gases = ·003665.
This volume of air would weigh

$$\frac{V}{1 + at} \times \frac{p - f}{760} \times \frac{·001293}{1} \text{ grms.};$$

for 1 c.c. of air at 0° and 760 mm. weighs ·001293 grms.

Now assuming that ether has the same coefficient of expansion as air, and that it could exist in the gaseous state at

0° and 760 mm., it is obvious that $w$ grms. thereof under these conditions would occupy

$$\frac{V}{1 + \cdot003665t} \times \frac{p-f}{760} \text{ c.c.}$$

But the specific gravity of a vapour, $s$, is the ratio of the weight of a certain volume thereof to the weight of the same volume of air measured under the same conditions. Hence we

have
$$s = \frac{w}{\dfrac{V}{1 + \cdot003665t} \times \dfrac{p-f}{760} \times \dfrac{\cdot001293}{1}}.$$

[It will be noted that the value found for $s$ is independent of the temperature of the bath ; it matters not what this may be so long as it is high enough to rapidly vaporise the body experimented on, and is constant during the experiment.]

The theoretical specific gravity of ether vapour (air = 1) is 2·56.

It is generally necessary to purify the ether. This is done by allowing it to stand for several hours over fused $CaCl_2$; then pouring it off into a flask containing thin slices of sodium. The flask should be provided with a cork through which an open chloride of calcium tube is inserted ; this allows of the escape of hydrogen, but prevents absorption of atmospheric moisture by the ether. When the hydrogen bubbles have entirely ceased to be disengaged, the ether is anhydrous and free from alcohol, and is ready for use.

(2) The specific gravity of alcohol vapour. Since alcohol has a much higher boiling point than ether [alcohol, B. P. = 78°; ether, B. P. = 35·5°] it is advisable in determining its specific gravity to use an aniline-vapour bath instead of a steam bath. Aniline boils at 182°. At this temperature alcohol is *rapidly* vaporised ; whereas at a temperature of 100° the vaporisation although complete would be much slower. But the more rapid the vaporisation the less chance is there of error due to the diffusion of the vapour into the colder parts of the apparatus and its condensation there.

The alcohol used should first be rendered as nearly an-hydrous as possible by letting it stand for several hours in a corked flask either over *anhydrous copper sulphate* or over freshly-burnt and coarsely powdered *quicklime*, and finally distilling off on a water bath.

The experiment is executed, and the result calculated, exactly as in the case of ether.

(3) The specific gravity of water vapour. Although an aniline bath would serve in this case, yet it is well that the student should familiarise himself with the use of the sulphur bath, which is much used when a high and constant temperature is desired. Coarsely powdered sulphur is at first cautiously heated in the bath $A$ (Fig. 43) till it melts, and then strongly heated till it boils, which it does at a temperature of about $440^{\circ}$. The experiment is conducted, and the result calculated, as before.

You now know that the molecular weight of water-gas is approximately 18. To find whether it is exactly 18 or not, you must determine the mass of oxygen which combines with 1 part by weight of hydrogen to form water.

**Exp. 2.** *Quantitative synthesis of water.* Arrange an apparatus as shewn in Fig. 44. $A$ is a piece of hard glass

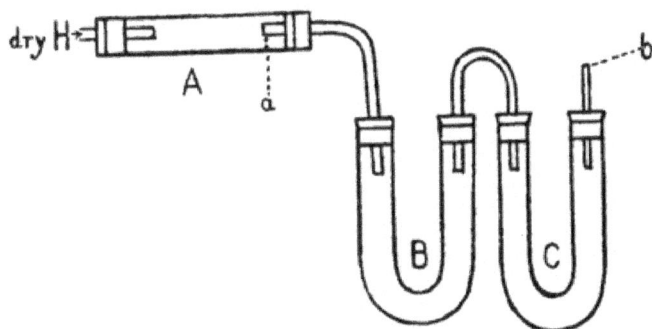

Fig. 44.

tubing about 15—20 centims. long; $B$ and $C$ are small light U tubes containing dry calcium chloride; each tube is thoroughly dried before the calcium chloride is put into it; the corks of $B$ and $C$ must fit very securely; these corks are dried in a steam bath before use. Two little pieces of caoutchouc tubing are provided, each fitted at one end with a small glass rod as a stopper, of such a bore that they will just pass over the ends of the tubes $a$ and $b$. The tube $A$ is thoroughly dried. A quantity of *pure copper oxide*, CuO (prepared from copper wire), is strongly heated over a lamp, and then quickly transferred to $A$, which should be about $\frac{3}{4}$ filled with it; a good cork carrying a tube of calcium chloride is at once placed in one end of $A$, the other end is firmly closed by a good cork, and $A$ is allowed to cool. The ∪ tubes are closed by the caoutchouc caps placed over $a$ and $b$. A gas-holder is now filled with hydrogen, prepared by the interaction of pure zinc and pure sulphuric acid diluted with about 20 volumes of water in a flask kept cold by being surrounded with cold water, the acid being added when it is quite cold and then in small successive quantities. The exit tube of the gas-holder is connected with 2 or 3 bottles containing concentrated sulphuric acid, and these are followed by a pair of large ∪ tubes containing calcium chloride and a small tube containing phosphorus pentoxide. The hydrogen is caused to flow in a gentle stream through the drying bottles and tubes for a little time. When $A$ is cold, it is carefully weighed; the tubes $B$ and $C$ are also weighed. The caoutchouc caps are withdrawn from the tubes $a$ and $b$ (the caps must be carefully preserved); the tube $a$ is pushed through a good cork which is then tightly fitted into $A$. A very good, soft, tightly fitting, cork, carrying a short piece of glass tubing, is now placed in the other end of $A$, and this glass tubing is connected with the hydrogen apparatus. A slow stream of hydrogen is allowed to pass through the entire apparatus for about 5 or 10 minutes; the tube $A$ is then cautiously heated beginning at the end nearest the calcium chloride tubes and gradually extending to the other end. It is advisable to have two Bunsen-lamps, so that while one is used in heating the main part of $A$ the other may keep the end of $A$ nearest the ∪ tubes so hot that water cannot condense there. To prevent all risk of charring the cork, it is well to have a disc of asbestos cloth with a slit in it which can be slipped over the tube near the cork. The heating of $A$ is continued until the colour shews that the greater part of the black copper oxide has been converted into red copper. Before this point is reached an apparatus for evolving a stream of carbon dioxide is prepared. The hydrogen-generator is removed, and the exit tube from the flask in which carbon dioxide is being prepared is attached to the drying bottles and tubes. The tube $A$ is kept red hot for 10 or 15 minutes while the carbon dioxide stream passes through it; by this means the last traces of water are driven out of $A$ into $B$ and $C$. The carbon dioxide is then stopped, and the apparatus is allowed to cool. The parts are detached, and the corks &c. are adjusted as at the beginning of the experiment; $A$ is weighed, and also $B$ and $C$ together. The decrease in the weight of $A$ gives the weight of oxygen which has combined with hydrogen to form water ; the increase in the weight of $B$ and $C$ gives the water formed ; the difference between the weight of water and that of oxygen gives the weight of hydrogen which has combined with the oxygen to form water.

8 parts by weight of oxygen combine with 1 part of hydrogen to form 9 parts of water.

The results of Exps. 1 and 2 in this Chap. shew that 16 parts by weight of oxygen combine with 2 parts by weight of hydrogen to form a molecule of water-gas. The atomic weight of oxygen is therefore not greater than 16, but it may be less than this number.   In order to determine the value of the atomic weight of oxygen it would be necessary to find the composition of a number of compounds of oxygen and to determine the molecular weight of each (s. "ELEMENTARY CHEMISTRY," Chap. XV., par. 297). The following experiments (3 and 4) are performed with the object of finding whether the atomic weight of oxygen is 16 or a submultiple of 16 ; they also serve to establish a probable value for the atomic weight of carbon.

Exp. 3.   *Determine the molecular weight and composition of carbon dioxide.*   Arrange an apparatus as shewn in Fig. 45.

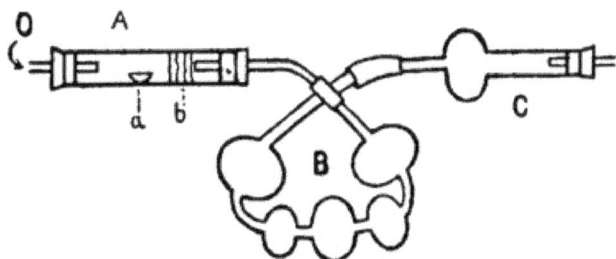

Fig. 45.

$A$ is a piece of hard glass tubing about 15—20 centims. long, communicating with the potash bulbs $B$, which again are in connection with the small light tube $C$ ; at $b$ is a little *dry* pure *copper oxide* (to oxidise any carbon monoxide that might be produced in the combustion).   The bulbs $B$ are filled with a solution of potash (1 part KOH in 2 parts water) to such an extent that a fairly rapid stream of gas may be sent through the bulbs without danger of any liquid being lost by spirting. Tube $C$ contains loosely packed solid potash.   The other end of $A$ communicates with an apparatus for supplying a regulated stream of pure dry *oxygen*. The oxygen is stored in a gas-holder, from which it is passed through two bottles of potash solution to absorb carbon dioxide, and then through two bottles of concentrated sulphuric acid and a large U tube containing calcium chloride to absorb moisture. The bulbs $B$ are closed by caoutchouc caps and weighed; the ends of $C$ are also closed and the tube is weighed. A small porcelain boat (indicated by $a$ in the fig.) is strongly heated, allowed to cool, and weighed; a little bit of *diamond* is then placed in the boat which is again weighed. The boat is now placed in the tube $A$ and the apparatus is arranged as shewn in the figure. The copper oxide, $b$, is gradually raised to full redness, a slow stream of oxygen is passed through the apparatus, and the diamond is heated until it burns in the oxygen. The product of the burning is carbon dioxide, which is absorbed by the potash in $B$ and $C$. When the burning is finished the apparatus is allowed to cool; $B$ and $C$ are weighed together, with the caoutchouc caps on; and the boat is withdrawn and weighed. A very little ash remains when diamond is burnt; the weight of ash found is deducted from the weight of diamond used, and the difference gives the weight of pure carbon burnt to carbon dioxide. The increase in the weight of $B$ and $C$ gives the weight of carbon dioxide formed. Assuming that the sole product of the burning was carbon dioxide, the results obtained enable the composition of this compound to be established.

The results of this experiment, if very carefully conducted, shew that 8 parts by weight of oxygen combine with 3 parts by weight of carbon to form 11 parts by weight of carbon dioxide. The molecular weight of carbon dioxide is therefore $n11$, where $n$ is a whole number. We must now determine the value of $n$. An approximately accurate determination of the specific gravity of the gas carbon dioxide will evidently suffice for this purpose.

Clean and dry a light flask of about 250 c.c. capacity. Close the flask by a very good well-fitting cork fitted with a short exit tube closed by a piece of caoutchouc tubing and a screw-clamp, and weigh it. In weighing the flask it is advisable to place another similar flask in the other pan and then to add weights until equilibrium is established. Note the temperature and pressure of the air. Pass a fairly rapid stream of pure, *perfectly dry, carbon dioxide* into the flask by a tube reaching to the bottom; after 10 or 15 minutes very slowly withdraw this tube and instantly cork the flask, taking care that the cork is pushed into the neck to the same distance as when the first weighing was executed.   Open the screw-clamp for a moment to establish equilibrium of pressure within and without the flask, and again at once close the clamp.   Weigh again.   Open the flask under *potash* solution (1 of potash in 2 of water); if the potash does not rush in and entirely fill the flask (shewing that the flask contained only carbon dioxide) begin the Exp. again.   Make a mark on the neck of the flask at the point to which the cork reaches.   Remove the cork; fill the flask to the mark on the neck with water and determine the volume of this water; let it be $a$ c.c.   You have thus determined the capacity of the flask.   Calculate the volume which the $a$ c.c. of air at the observed temperature and pressure will occupy at $0°$ and 760 mm.; then, knowing that 1000 c.c. of air at $0°$ and 760 mm. weigh 1·293 grams, find the weight in grams of this volume of air; that is, find the weight of air in the flask when it was weighed full of air.   Now deduct this weight from the observed weight of flask and air; the difference is the weight of the empty flask.   Deduct the weight of the empty flask from that of the flask filled with carbon dioxide; the difference is the weight of the carbon dioxide.

You now know the weights of equal volumes ($a$ c.c.) of air and carbon dioxide at the same temperature and pressure. From these data find the specific gravity of carbon dioxide referred to air; multiply this number by 28·87, and you obtain the specific gravity of the gas referred to hydrogen as twice unity, that is you obtain an approximately accurate value for the molecular weight of carbon dioxide.

The result of this Exp. shews that the molecular weight of carbon dioxide is about 44.   But you have already found that 3 parts by weight of carbon combine with 8 parts by weight of oxygen, and that the molecular weight of carbon dioxide is therefore $n11$; you have now found the value of $n$ to be 4.

As 8 parts by weight of oxygen combine with 3 parts of carbon, there must be 32 parts by weight of oxygen combined with 12 parts of carbon in 44 parts by weight, that is in a molecule, of carbon dioxide.

The results of Exp. 2 shewed that the atomic weight of oxygen is not greater, although it may be less, than 16; the results of the present Exp. shew that the atomic weight of oxygen is probably not less than 16.

The results of the present Exp. also shew that as 12 parts by weight of carbon combine with oxygen to form a molecule of carbon dioxide, the atomic weight of carbon is not greater than 12.

**Exp. 4.** *Determine the composition and molecular weight of carbon monoxide; assuming the specific gravity of the gas to be known.*

Carbon monoxide gas is about 14 times heavier than hydrogen; therefore its molecular weight is approximately 28.

Fill a eudiometer with clean and dry mercury and invert it in a mercury-trough. Arrange a small flask with a funnel tube, and an exit tube connected with a bottle containing potash solution to absorb any carbon dioxide produced in the reaction; let the exit tube from this bottle be somewhat narrowed at the open end and bent so that this end can be easily brought under the open end of the eudiometer in the trough. Place about 20 c.c. of concentrated *formic acid* solution and about 100 c.c. of concentrated *sulphuric acid* in the flask, and gently heat until a stream of carbon monoxide passes through the drying bottles ($H_2CO_2 - H_2O = CO$).

*As carbon monoxide is very poisonous these experiments must be conducted in a draught place.*

After the lapse of 10 minutes or so you may be sure that the issuing gas is free from air. Now allow about 30 c.c. of the pure carbon monoxide to pass into the eudiometer; withdraw the lamp; and set aside the eudiometer. Meanwhile put about 5 grams of *pure dry potassium chlorate* in a tube of hard glass; draw out the open end of the tube and bend it into the shape of the delivery tube of an apparatus for making and collecting oxygen or hydrogen. Place this tube in a clamp, and heat it gradually until a rapid stream of oxygen is evolved. While this is proceeding, read off the level of the mercury in the eudiometer, also read the thermometer and barometer, and the height of the column of mercury in the eudiometer, and so determine the volume of carbon monoxide. By this time pure oxygen, free from air, will be issuing from the tube in which the potassium chlorate is heated; cause the oxygen to be evolved rapidly, then remove the lamp, and quickly carry the tube to the eudiometer into which pass about 80 to 100 c.c. of oxygen. Press down the eudiometer firmly on a pad of india-rubber moistened with a solution of

mercuric chloride, and clamp it; then pass an electric spark
through the gases; the carbon monoxide is burnt to dioxide
$(CO + O = CO_2)$. Now slowly remove the eudiometer from the
pad and allow the mercury to rise. Read the level of the
mercury in the eudiometer, also the height of the column of
mercury in the eudiometer; these give the volume of carbon
dioxide *plus* oxygen now in the eudiometer.

Bring a few c.c. of a solution of *potash* (1 part potash in 2
of water) into the eudiometer by means of a bent pipette.
The carbon dioxide is absorbed; when absorption is complete,
read the level of the liquid in the eudiometer, and the height
of the column of mercury; the contraction which has occurred,
making allowance for the short column of potash solution the
specific gravity of which may be taken as 1·32, measures the
carbon dioxide in the tube when the potash was added.

You have now determined that a certain volume of wet
carbon monoxide, say $a$ c.c., at a specified temperature and
pressure, when burnt gave a certain volume, say $b$ c.c., of wet
carbon dioxide at a specified temperature and pressure. The
volumes of the dry gases at $0^\circ$ and 760 mm. must now be
calculated; then, knowing that 1000 c.c. of carbon monoxide
at $0^\circ$ and 760 mm. weigh 1·251 grams, and 1000 c.c. of carbon
dioxide at $0^\circ$ and 760 mm. weigh 1·966 grams, it is easy to
find the weights of the volumes of the two oxides.

When this calculation is made, you have determined that a
certain weight, $a$ grams, of carbon monoxide when burnt gave
a certain weight, $b$ grams, of carbon dioxide; as you already
know the gravimetric composition of carbon dioxide it is easy
to find the weight of carbon in these $b$ grams; but this is also
the weight of carbon in the $a$ grams of carbon monoxide; hence,
as this oxide is a compound of carbon and oxygen only, you
have the data for finding the composition of carbon monoxide.

The result of this experiment, if carefully conducted, is
that 8 parts by weight of oxygen are combined with 6 parts of
carbon in 14 parts of carbon monoxide.

The molecular weight of the gas is therefore $n14$ where $n$
is a whole number; but $n$ must be equal to 2, because carbon
monoxide is about 14 times heavier than hydrogen.

A molecule of carbon monoxide is therefore composed of
16 parts by weight of oxygen combined with 12 parts by
weight of carbon.

The results of this experiment confirm the numbers 16 and
12 as the atomic weights of oxygen and carbon respectively.

# CHAPTER IV.

## DISSOCIATION.

**Exp. 1.** DETERMINE the specific gravity (referred to hydrogen as unity) of the vapour obtained by heating *ammonium chloride*. For this purpose use v. Meyer's apparatus heated in a bath of molten lead to a temperature of about 500° or so. The bath may consist of a piece of iron gas-pipe about $2\frac{1}{2}$ inches wide and 8 or 10 inches long closed at one end by an iron plug screwed on. The lower end of the vapour density apparatus is rested on the surface of the solid lead; the lead is gradually melted by means of a large Bunsen-lamp (lead melts at about 365°), and the apparatus is allowed slowly to sink into the molten metal.

Assuming the atomic weights of nitrogen and chlorine to be 14 and 35·5 respectively, the simplest formula that can be given to ammonium chloride consistently with its composition is $NH_4Cl$. If this formula represents the composition of a molecule of the compound in the state of gas, the molecular weight of gaseous ammonium chloride must be 53·5, and the gas must be about 26 times heavier than hydrogen

$$\left(\frac{14 + 4 + 35\cdot5}{2} = 26\cdot75\right).$$

But the result of your experiment shews that the vapour obtained by heating ammonium chloride is about 13 times heavier than hydrogen. You would therefore be inclined to deduce the number 26 as the approximate value of the molecular weight of ammonium chloride gas; but if this value is correct, either the formula $N_{\frac{1}{2}}H_2Cl_{\frac{1}{2}}$ must be given to the compound, or the accepted atomic weights of nitrogen and

chlorine must be halved.    Let us further examine the action
of heat on ammonium chloride.

Exp. 2.    Arrange an apparatus as shewn in Fig. 46.    $A$ is
a piece of hard glass tubing about 30 centims. long and 2 or $2\frac{1}{2}$

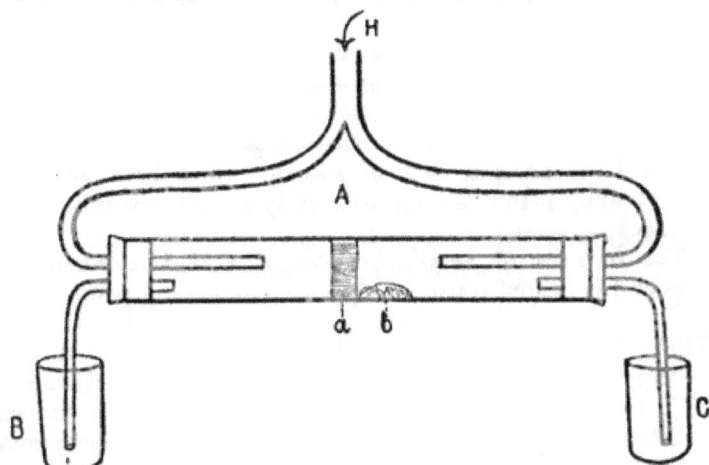

Fig. 46.

centims. wide; $a$ is a dry plug of plaster of Paris or of fairly
tightly packed asbestos; a little freshly sublimed *ammonium
chloride* is placed at $b$.    The beakers $B$ and $C$ contain water;
a few drops of red litmus are poured into $B$, and a few drops
of blue litmus into $C$.    A rapid stream of *hydrogen* is passed
into the apparatus as indicated by the arrow : when the air
is all driven out of the apparatus the ammonium chloride at $b$
is heated, the hydrogen-stream being continued.    The litmus
in $B$ soon turns blue, and that in $C$ turns red.    This indicates
that the hydrogen passing out of the apparatus into $B$ carries
an alkali with it, and that passing into $C$ carries an acid with
it.    But the only acid and alkali that could be produced under
the conditions of the experiment are hydrochloric acid and
ammonia, respectively.

Part of the ammonium chloride is dissociated in this Exp.
by the heat into ammonia and hydrochloric acid; the ammonia,
being about half as light, bulk for bulk, as the hydrochloric
acid gas, diffuses more rapidly than the acid through the plug
$a$, the hydrogen issuing into $B$ therefore carries with it some
free ammonia ; for the same reason, the hydrogen issuing into
$C$ carries with it some free hydrochloric acid.

Further experiments on dissociation will be found in Part
III. (Chap. II.).    Meanwhile an experiment may be performed
to illustrate the general statement that the amount of dissociation of a dissociable compound increases as temperature
increases (up to a certain limit), pressure being kept constant.

**Exp. 3.**    Make 3 determinations of the specific gravity of
the vapour obtained by heating *amylic bromide* by V. Meyer's
method, (1) using a bath of *methyl salicylate* which boils at
222°, (2) using a bath of *bromonaphthalene* which boils at
280°, and (3) using a bath of *sulphur* which boils at 450°.
(These boiling points are approximately correct, pressure being
760 mm.)

The formula of amylic bromide is $C_5H_{11}Br$; when the
compound is heated considerably above its boiling point (129°)
it is dissociated into amylene ($C_5H_{10}$) and hydrobromic acid
(HBr); dissociation begins at about 160°, and at about 360°
the vapour consists wholly of a mixture in equal volumes of
$C_5H_{10}$ and HBr; at intermediate temperatures the vapour
consists of $C_5H_{11}Br$, mixed with $C_5H_{10}$ and HBr.    The specific
gravity of gaseous $C_5H_{11}Br$, referred to air as unity, is 5·22; the
specific gravity of a mixture of equal volumes of $C_5H_{10}$ and
HBr gases is 2·61.    If no dissociation occurs at a specified
temperature the observed specific gravity will be 5·22; if
complete dissociation occurs the observed specific gravity will
be 2·61; and if partial dissociation occurs the observed
specific gravity will be a number between 5·22 and 2·61.

The vapour obtained by heating amylic bromide to about 220° has a
specific gravity of about 4·2; at 280° the specific gravity is about 3·1; and
at 450° it is 2·6; pressure in each case being normal.

*Reference to* "ELEMENTARY CHEMISTRY."    Chap. XII. pars.
233 to 236; also Chap. XVI. pars. 333 to 337.

# CHAPTER V.

## REACTING WEIGHTS OF COMPOUNDS DETERMINED BY CHEMICAL METHODS.

THE relative weight of the *smallest mass of a compound which takes part in chemical interactions* is determined by quantitatively studying some of the typical reactions of the compound.

**Exp. 1.** *Determine the reacting weight of acetic acid* (a compound of H, C, and O); *assuming the acid to be monobasic, and assuming the atomic weight of silver to be* 108.

Silver acetate is decomposed by heating in air; silver remains, and the carbon hydrogen and oxygen are burnt away. Prepare pure *silver acetate* by adding a solution of recrystallised *sodium acetate* to a concentrated solution of *silver nitrate*, collecting the crystals which form, and crystallising once or twice from hot water. Dry the crystals of silver acetate by pressing between filter paper and then heating to 100°. Weigh out about ·5 gram of the dry salt into a weighed porcelain crucible with a lid, then heat gently to full redness over a Bunsen-lamp, and then for a few minutes over the blowpipe (keeping the lid on the crucible), cool, and weigh the silver in the crucible.

Calculate the percentage of silver, and deduct this from 100 ; the difference gives the carbon, hydrogen, and oxygen, combined with the silver to form 100 parts of silver acetate. The atomic weight of silver is 108 ; calculate the mass of carbon, hydrogen, and oxygen, combined with 108 parts by weight of silver to form silver acetate. As acetic acid is monobasic, the acid would be obtained from the silver salt by replacing 108 of silver by 1 of hydrogen: make the necessary calculation, and thus find the reacting weight of acetic acid.

**Exp. 2.**  *Assuming the composition of a reacting weight of oxalic acid to be represented by the formula* $nHCO_2$ *(where* n = *a whole number), find the probable value of* n.

Prepare a considerable quantity of a fairly concentrated solution of *potash*; about 150 grams potash in 300 c.c. water. Dissolve in water about 60 grams of *oxalic acid*; divide the solution into three equal parts; exactly neutralise one part by the potash solution, and to another part add half as much of the same potash solution; evaporate both solutions to the crystallising point; collect the crystals which form, and recrystallise them twice from hot water. Dry each set of crystals by pressure between paper, and then at $100°$; weigh out equal quantities of the two solids, add *sulphuric acid* to each, warm, and run in a standardised solution of *potassium permanganate* until a faint pink colour is permanently produced.

The fact that a certain mass of one of the solids requires a quantity of permanganate to completely oxidise it very different from that required by an equal mass of the other solid, shews that the two sets of crystals are different salts. (It is assumed that oxalates are wholly oxidised by permanganate in presence of sulphuric acid.) Hence at least two potassium salts can be obtained by interactions between potash and oxalic acid: hence the composition of a reacting weight of oxalic acid is expressed by a formula not less than $H_2C_2O_4$.

To the third quantity of oxalic acid (equal to those already used) add twice as much potash as you know is required for neutralisation; evaporate; collect the crystals which form; recrystallise them three or four times from hot water; dry; weigh out a quantity equal to that taken of the former crystals, and titrate with permanganate as before. The result shews that these crystals are the same salt as was produced by exactly neutralising oxalic acid with potash.

So far as these experiments go, oxalic acid appears to form two, and not more than two, potassium salts; hence you conclude that the acid is dibasic; and hence that the probable value of $n$ in the formula $nHCO_2$ is two, that is, the composition of the reacting weight of oxalic acid is probably expressed by the formula $H_2C_2O_4$.

**Exp. 3.**  *Find the reacting weight of barium oxide, by examining the reaction which occurs between this base and an acid the reacting weight of which is known.*

It will be necessary to determine the simplest formula to be given to barium oxide, assuming the atomic weights of barium and oxygen to be 137 and 16 respectively, and knowing that barium and oxygen are combined in barium oxide in the ratio 137 : 16.

Weigh out accurately about ·5 gram of *barium oxide*; add a measured quantity of dilute *sulphuric acid* the strength of which is known, adding so much that the liquid above the solid barium sulphate which forms is strongly acid to litmus; dilute; warm for some time; then allow the pp. of barium sulphate to settle; add a drop or two of litmus solution, and determine the amount of sulphuric acid which remains unchanged by means of a standardised alkali solution.

Calculate the mass of sulphuric acid which has been used to convert the barium oxide into sulphate; then calculate the mass of barium oxide which has interacted with each reacting weight ($H_2SO_4 = 98$) of sulphuric acid.

The result is that 98 parts by weight of sulphuric acid react with 153 parts of barium oxide; but 98 is the reacting weight of sulphuric acid; therefore the reacting weight of barium oxide is probably not greater than 153. The formula BaO ($Ba = 137$, $O = 16$) expresses the composition of 153 parts by weight of barium oxide; therefore the formula BaO probably expresses the composition of the reacting weight of barium oxide: the experiment with sulphuric acid shews that the reacting weight is probably not greater than 153; but it cannot be less than this if the atomic weights of barium and oxygen are 137 and 16, respectively.

Supplement the result obtained by performing a similar experiment, using standardised *hydrochloric acid* instead of sulphuric acid. In this case the barium chloride formed remains in solution; after adding a decided excess of hydrochloric acid, determine the acid which remains by means of a standardised *ammonia* solution. Assuming the reacting weight of hydrochloric acid to be 36·5 (HCl), the results of the experiment shew that one reacting weight of this acid interacts with $\frac{153}{2} = 76\cdot5$ parts by weight of barium oxide. But the reacting weight of barium oxide cannot be less than 153; therefore we conclude that this oxide and hydrochloric acid interact to form barium chloride in the ratio, expressed in reacting weights, of 1 : 2. This result confirms that obtained by using sulphuric acid.

**Exp. 4.** *Find the reacting weight of aniline ; assuming that the double compound which this base forms with hydrochloric acid and platinic chloride has the composition* $2XHCl . PtCl_4$, *where* $X = one\ reacting\ weight\ of\ aniline.*

It is first necessary to purify the '*pure*' *aniline* of commerce as follows. Equal masses (say 25 grams) of *aniline* and *glacial acetic acid* are mixed in a retort which is tilted upwards and attached to a long glass tube which serves as a condenser. The contents of the retort are heated to boiling for 12 hours, after which time the formation of acetanilide may be presumed to be complete. While the retort is still hot it is connected with a condenser, and heating is continued until the liquid which passes over begins to solidify. This will happen at about $280°$—$290°$. The condenser is now removed and the receiver changed. The impure acetanilide which collects in the receiver is purified by recrystallising once or twice from a large quantity of boiling water. The acetanilide is reconverted into aniline by heating it with potash in a flask fitted with an inverted condenser, and the aniline is blown over with steam. The aniline is separated from the water as completely as possible, dehydrated by means of solid potash, and finally distilled ; the portion boiling at $181°$—$182°$ is pure aniline. To this pure aniline is added a slight excess of hydrochloric acid, and the solution is evaporated to dryness at $100°$; the residue is dissolved in a small quantity of water, and the liquid is filtered if necessary. To the filtrate a fairly strong solution of *platinic chloride* is added ; *but care must be taken that the platinum salt is not in excess.* The double chloride of aniline and platinum which is precipitated is strained off from the mother liquor, and a small quantity is dried at $100°$ in a weighed crucible. The crucible *plus* dried salt is weighed; and the salt is then decomposed by heating it at first gently and finally strongly over a Bunsen-flame. The residual metallic platinum is weighed.

Let $w'$ = weight of double chloride

,, $w''$ = ,, ,, residual platinum,

and let X represent the reacting weight of aniline. Since the platinum double salts of nitrogenous bases are constituted on the same type as the platinum double salts of ammonia, it follows that $2XHCl . PtCl_4$ represents the composition of the aniline platinum chloride.   The reacting weight of this double chloride is $2X + 73 + 339\cdot5$ ; on decomposition this mass would yield $197\cdot5$ of platinum.   Hence the equation must hold

$$\frac{2X + 73 + 339\cdot5}{197\cdot5} = \frac{w'}{w''}$$

and from this X is easily determined.

It is to be observed that this is a fairly general method for finding the reacting weights of organic bases.

**Exp. 5.**   *The reacting weight of iron-ammonium alum is expressed by the formula* $n(NH_4Fe(SO_4)_2.12H_2O)$; *find whether the value of* n *is* 1 *or* 2.

If the formula which expresses the composition of a reacting weight of iron alum is $(NH_4)_2Fe_2(SO_4)_4.24H_2O$ it may be possible to partially dehydrate this compound and obtain the hydrate $(NH_4)_2Fe_2(SO_4)_4.H_2O$, whereas if the formula of the alum is $NH_4Fe(SO_4)_2.12H_2O$ the composition of the lowest hydrate which could exist would be $NH_4Fe(SO_4)_2.H_2O$ $= (NH_4)_2Fe_2(SO_4)_4.2H_2O$.

Determinations of the ratio of the water lost to the solid remaining when iron alum is heated to different temperatures will therefore enable us to decide between the two possible formulae for this compound, provided it can be proved that dehydration is the only change which occurs when the alum is heated to these temperatures.

[Before beginning the Exp. read L u p t o n on *The formulae of the alums*, C. S. J o u r n a l, (2). 13, 201.]

Recrystallise some *iron-ammonium alum* from water; powder a few grams, dry by pressure between paper, and at once weigh out about ·5 gram into a small porcelain boat.   Place the boat in a small hard glass tube ; connect one end of the tube with a couple of drying bottles containing concentrated sulphuric acid and the other with bulbs containing *Nessler's reagent* for detecting ammonia.   Arrange the tube in a small air bath fitted with a thermometer ; draw a slow current of air through the tube, and heat to 150° until the weight of the substance in the boat is constant.   Observe that ammonia is not given off during the operation.   Then raise the temperature to 230° until no further loss of weight occurs, observing that ammonia is not evolved.

Having thus proved that the changes which occur at 150° and 230° are simple dehydrations, calculate the composition of the salt which remained after heating to 150° and of that which remained after heating to 230°. Finally prove that moist ammonium sulphate evolves ammonia at about 120°—125°.

Shew that your results confirm the formula

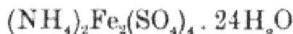

$$(NH_4)_2Fe_2(SO_4)_4 . 24H_2O$$

for iron-ammonium alum.

The salt formed at 150° is $(NH_4)_2Fe_2(SO_4)_4.H_2O$, and at 230° this is completely dehydrated.

*Reference to* "ELEMENTARY CHEMISTRY." Chap. VI. pars. 84 to 86; also Chap. XI. par. 163; also Chap. XVI. par. 315.

# CHAPTER VI.

## CHEMICAL CHANGE.

In Part I. Chap. XVI. experiments were conducted to illustrate some of the conditions which chiefly affect the progress and final results of chemical reactions. Exp. 10 in that Chapter was conducted by causing solutions of ferric chloride and potassium sulphocyanide to react under conditions which were all constant except the relative masses of the reacting bodies; the amount of change increased as the mass of one of the reacting bodies increased relatively to that of the other.

The following experiment is performed with the same two compounds, ferric chloride and potassium sulphocyanide, but it is made quantitative.

**Exp. 1.** Prepare pure *ferric chloride* by passing a rapid stream of pure dry *chlorine* over pure *iron wire* heated to redness; resublime the chloride from a small tube of hard glass into a small bottle with a good stopper.

Prepare an aqueous solution of this ferric chloride of known strength; about 3 grams per litre is a convenient strength.

Experiments made by Gladstone shewed that when ferric chloride and potassium sulphocyanide interact in about the ratio $Fe_2Cl_6$ : 500KCNS the whole, or nearly the whole, of the ferric chloride is changed to ferric sulphocyanide. On the basis of this experimental result, prepare a standard solution of *ferric sulphocyanide*, by adding to a measured volume of the ferric chloride solution a quantity of potassium sulphocyanide such that the two salts are in the ratio $Fe_2Cl_6$ : 500KCNS. A convenient strength to make this standard solution is about ·03 gram $Fe_2(CNS)_6$ in 100 c.c.

Prepare a solution of *potassium sulphocyanide*, by dissolving a weighed quantity of the pure salt in water, of a strength such that equal volumes of this solution and of the solution of ferric chloride contain KCNS and $Fe_2Cl_6$ in about the ratio 10 : 1.

Now place a measured quantity, say 20 c.c., of the standard ferric sulphocyanide solution in a cylinder of white glass; into another exactly similar cylinder measure out small equal volumes of the ferric chloride and potassium sulphocyanide solutions, say 2 c.c. of each, the volume of ferric chloride solution containing exactly the same mass of ferric chloride as was used to prepare the volume of standard ferric sulphocyanide solution in the other cylinder.

Now add exactly enough water to make the total volume of the liquid equal to that of the standard solution in the other cylinder. The colour of the standard ferric sulphocyanide should now be considerably darker than that of the mixed solutions. Now run water from a burette into the standard solution until the depth of colour is the same in the two cylinders; measure the volume of water required.

Perform similar experiments, using in each the same volume of the ferric chloride solution, but increasing that of the potassium sulphocyanide solution; the ratio of KCNS : $Fe_2Cl_6$ being (say) 10 : 1, 15 : 1, 20 : 1, 40 : 1, 80 : 1, 100 : 1, and 200 : 1. Calculate, in each case, (1) the mass of ferric sulphocyanide which would be obtained from the ferric chloride used supposing the whole of it were changed to ferric sulphocyanide; and (2) the mass of ferric sulphocyanide actually obtained. The second calculation is made by measuring the water added to the standard solution to make the depth of colour the same as that of the other solution, and then saying, as the total volume of the standard liquid is to the volume of this liquid before dilution, so is the mass of ferric sulphocyanide which would be formed if the whole of the iron salt had been changed to the mass of ferric sulphocyanide actually formed.

Arrange your results to shew the ratio of KCNS to $Fe_2Cl_6$ in each experiment, and the percentage of possible $Fe_2(CNS)_6$ actually produced.

The following examples are given to shew the kind of results obtained. 1·8194 grams $Fe_2Cl_6$ were dissolved in 500 c.c. water: to 50 c.c. of this solution 27·27 grams KCNS were added and whole was diluted to 800 c.c. (this is called *standard ferric sulphocyanide solution*, ratio of KCNS to $Fe_2Cl_6$ in this solution $=500 : 1$). $4\cdot586$ grams KCNS were dissolved in 211 c.c. water, equal volumes of this solution and of the ferric chloride solution contained KCNS and $Fe_2Cl_6$ in ratio 20 : 1.

I.  20 c.c. *standard ferric sulphocyanide solution* (to form which $\cdot00455$ grm. $Fe_2Cl_6$ were used).

$1\cdot25$ c.c. ferric chloride solution (containing $\cdot00455$ grm. $Fe_2Cl_6$) $+1\cdot25$ c.c. potassium sulphocyanide solution $+17\cdot5$ c.c. water $(=20$ c.c. in all).

18 c.c. water were added to standard solution to make depths of colour equal.

Now $1\cdot25$ c.c. of the ferric chloride solution contain $\dfrac{5}{200} \times \cdot18194$ grm.

$Fe_2Cl_6$; if this were all changed to $Fe_2(CNS)_6$ it would give $\dfrac{5}{200} \times \cdot18194 \times \dfrac{460}{325}$ $= \cdot006437$ grm. $Fe_2(CNS)_6$. But $\cdot003388$ grm. $Fe_2(CNS)_6$ was actually formed ($38 : 20 = \cdot006437 : \cdot003388$). Therefore ratio of $Fe_2(CNS)_6$ formed to total possible $Fe_2(CNS)_6 = 52\cdot6 : 100$.

*Result.* $Fe_2Cl_6 :$ KCNS$=1 : 20$. Percentage of possible $Fe_2(CNS)_6$ actually formed $= 52\cdot6$.

II.  20 c.c. *standard ferric sulphocyanide solution.*

$1\cdot25$ c.c. ferric chloride solution $+2\cdot5$ c.c. potassium sulphocyanide solution $+16\cdot25$ c.c. water $(=20$ c.c. in all). 14 c.c. water required to equalise colours.

*Result.* $Fe_2Cl_6 :$ KCNS$=1 : 40$. Percentage of possible $Fe_2(CNS)_6$ actually formed $= 58\cdot82$.

The rates of many processes of chemical change vary with variations in the temperature at which the change occurs, other conditions being constant.

**Exp. 2.**  When *potassium permanganate* solution is slowly added to a hot acidified solution of *oxalic acid*, the latter is oxidised to carbon dioxide and water at the expense of a portion of the oxygen of the permanganate.

You are given solutions of *oxalic acid* ($H_2C_2O_4$) and *potassium permanganate* of stated strength. To a measured volume, say 50 c.c., of the oxalic acid solution add a considerable quantity of *sulphuric acid*, heat to $60^\circ$—$70^\circ$, and run in the permanganate from a burette until the oxidation is complete (until a slight pink colour is permanently produced in the liquid); note the volume of permanganate used.

Now measure out 4 equal quantities, say 50 c.c. each, of the oxalic acid solution, and add the same volume of the same moderately concentrated sulphuric acid to each. Surround one beaker by pounded ice; keep another at the temperature of the air; heat another to $30^\circ$—$35^\circ$ by plunging it into water placed over a low flame; and heat the fourth to $50^\circ$—$55^\circ$. Measure out 4 equal volumes of the permanganate solution into test tubes, the quantity in each case being that which you know, from the result of the preliminary experiment, just suffices to oxidise the oxalic acid in the beakers; arrange these test tubes alongside the oxalic acid solutions so that they are heated to the same temperatures. Also make about 50 c.c. of a fairly concentrated solution of *potassium iodide*, and a little *starch paste*; and obtain a standardised solution of *sodium thiosulphate*.

Observe the temperature of each oxalic acid solution; note the time accurately, and at once add the permanganate solution to each beaker by immersing the tube containing the permanganate in the liquid in the beaker; shake up once; after 1 minute add about 10 c.c. of the potassium iodide solution to each beaker. The permanganate which has not reacted with the oxalic acid interacts with the potassium iodide to produce iodine which remains in solution. Now bring all the beakers to the temperature of the air, and then determine the amount of free iodine by means of the standard sodium thiosulphate solution.

In order to find the value of the sodium thiosulphate solution in terms of the permanganate, perform an experiment similar to the foregoing, but using only permanganate, sulphuric acid, and potassium iodide. Find how many c.c. of the thiosulphate solution are required to react with the iodine set free from the 10 c.c. of potassium iodide solution by the action of that volume of the permanganate which was used in the experiments with oxalic acid.

From the data now obtained you can calculate the percentage of chemical change which has occurred at each of the four temperatures.

The amounts of some chemical changes are conditioned by the time during which the reactions proceed, other circumstances remaining the same.

**Exp. 3.** Arrange and conduct an Exp. similar to the last; keep the beakers at a low temperature by surrounding them with pounded ice; determine the quantity of unchanged *permanganate* at the end of 2, 5, 8, 10, and 20, minutes.

Arrange your results to shew the percentage of chemical change at the end of each period of time.

**Exp. 4.** When *ammonium carbamate* is brought into contact with *water* it is partially changed to ammonium carbonate; $CO . NH_2 . ONH_4 + H_2O + Aq = CO (ONH_4)_2 Aq$. The amount of change varies with variations in (1) the time, (2) the temperature, (3) the relative masses of the interacting substances. [Read Fenton, on *The limited hydration of ammonium carbamate*, Proc. R. S. 39, 386.]

When *ammonium carbamate* interacts with *sodium hypochlorite* (NaClO) in presence of soda, one-half of the nitrogen of the carbamate is evolved as nitrogen; this reaction presents a means for measuring the amount of change of carbamate to carbonate which has occurred under defined conditions.

The amount of carbamate remaining unchanged is measured as follows. A small flask is fitted with a caoutchouc cork and delivery tube dipping a very little under the surface of water in a shallow basin; a tube containing a strongly alkaline solution of sodium hypochlorite is placed inside the flask so that it rests against the side with its open end upwards; the solution containing ammonium carbamate is brought into the flask which is then corked, and the delivery tube is arranged so that it is quite full of air; a graduated tube filled with water is placed over the orifice of the delivery tube; the flask is surrounded by water to keep the temperature constant; it is then tilted so that the hypochlorite solution mixes with the carbamate solution, and the reaction is allowed to proceed, the flask being shaken from time to time, so long as gas collects in the graduated tube; when the reaction is complete the delivery tube is still just full of gas; the graduated tube is removed to a tall vessel filled with water, and the volume of nitrogen is read off with the usual precautions (*s.* Fenton, C. S. Journal, Trans. 1878, 300).

The interaction between ammonium carbamate and sodium hypochlorite is expressed thus in an equation;—

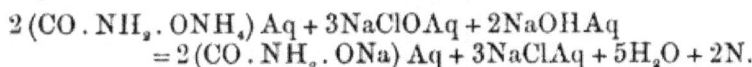

$$2 (CO . NH_2 . ONH_4) Aq + 3NaClOAq + 2NaOHAq$$
$$= 2 (CO . NH_2 . ONa) Aq + 3NaClAq + 5H_2O + 2N.$$

If $V =$ the total volume of nitrogen combined in a specified mass of ammonium carbamate, and $v =$ volume of nitrogen evolved by hypochlorite after partial change to carbonate, then $\dfrac{2v - V}{V}$ gives the ratio of the number of molecules of water which have reacted with the carbamate to the number of reacting weights of carbamate in the mass taken; this ratio may be used to conveniently express the amount of hydration which has occurred.

(1) **Influence of time.** Dissolve a weighed quantity of freshly prepared *ammonium carbamate* in water, and make up to a definite volume; say 6—7 grams in 100 c.c. Determine the amount of hydration which has occurred after definite intervals of time—say 10, 40, 100, and 200, minutes—by withdrawing measured quantities of the solution—say 5 c.c. in each case—and treating with sodium hypochlorite.

(2) **Influence of mass.** Dissolve 8—10 grams *ammonium carbamate* in 100 c.c. water; dilute 25 c.c. to 50 c.c. and another 25 c.c. to 250 c.c.; set aside in stoppered bottles for 4 days, and then determine the nitrogen evolved from 5 c.c. of the original solution, 10 c.c. of the more dilute, and 50 c.c. of the most dilute, solution (these volumes correspond to equal masses of carbamate originally taken).

(3) **Influence of temperature.** Repeat the experiments on the influence of time, conducting one series in vessels surrounded by ice, and another at the ordinary temperature.

*Reference to* "ELEMENTARY CHEMISTRY." Chap. XII.

Some elements exist in more than one form; such elements are said to exhibit *allotropy* and the change from one form to another is called *allotropic change.* Allotropic change is not accompanied by any change of mass.

**Exp. 5.** Fit a fair sized, wide, light, test tube with a good cork carrying short entrance and exit tubes, to each of which is attached a small piece of caoutchouc tubing passing through light screw clamps. Dry the apparatus, and arrange the tube as shewn in Fig. 47, so that it can be heated while a stream of *dry* carbon dioxide is passed through it.

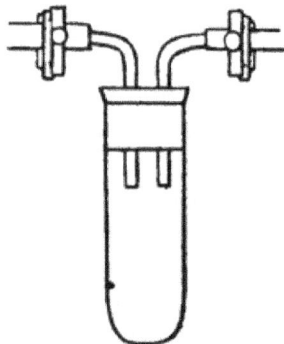

Fig. 47.

Pass *dry carbon dioxide* through the apparatus for a few minutes; then loosen the cork, drop into the tube a *dry* piece of *phosphorus* just cut from the middle of a stick, immediately replace the cork and continue the passage of the stream of carbon dioxide for 5 or 10 minutes. Now close the screw clamps firmly; detach the tube from the carbon dioxide apparatus, and weigh it. Replace the tube, and continue the carbon dioxide stream; surround the tube by dry sand in which a thermometer is placed; after a few minutes heat the sand, gradually raising the temperature to 220°—230°, and maintain the tube at this temperature, carbon dioxide passing *very slowly* through the apparatus all the time, until there seems to be no further change in the appearance of the contents of the tube. Allow to cool; close the screw clamps; detach the tube and weigh it. The weight is the same as it was at the beginning of the experiment. Compare the properties of the matter in the tube with those of ordinary phosphorus. Ordinary phosphorus is a soft, crystalline, very easily inflammable, solid; it is soluble in carbon disulphide: the matter in the tube is a red, amorphous, solid; it is insoluble in carbon disulphide, and is not easily inflammable. But the result of your experiment renders it very improbable that the solid in the tube can be a compound of phosphorus with other elements.

**Exp. 6.** Connect two short wide pieces of glass tubing each with an apparatus for supplying a stream of pure dry *oxygen*; bend the tubes to an obtuse angle and let the open end of each dip beneath the surface of 400 or 500 c.c. of water in a beaker. Place about ·5 gram (accurately weighed) of *red phosphorus* in one tube and about the same quantity of *ordinary phosphorus* in the other*; pass oxygen over each, and heat until the phosphorus is wholly burnt to oxide. Most of the oxide of phosphorus formed is swept into the water and there dissolved. When the combustions are completed detach the tubes from the oxygen apparatus, and place each in its own beaker; any traces of oxide of phosphorus in the tubes are thus dissolved in the water. Now remove the tubes; wash them thoroughly with cold water, and allow

---

* The ordinary phosphorus may be weighed in a very small tube which it nearly fills. Weigh this tube with a cork in it; cut a little bit of phosphorus, of a size to fill the tube, from the inside of a stick, instantly place it in the tube, insert the cork and weigh.

the washings to run into the beakers.   Boil the contents of
the beakers nearly to dryness, add water, and repeat this process
2 or 3 times.   Then estimate the phosphoric acid in each solution
by titration with a standardised solution of *uranium acetate*,
or by precipitation as $MgNH_4PO_4$ and weighing as $Mg_2P_2O_7$.
From the masses of phosphoric acid ($H_3PO_4$) found, calculate
the mass of phosphorus pentoxide ($P_2O_5$) produced by the
combustion of equal masses of the two varieties of phosphorus.
The result is that equal masses of the two kinds of phosphorus
produce the same mass of phosphorus pentoxide.

The results of this experiment and of Exp. 5 shew that
ordinary and red phosphorus are the same element.

Although the change from one variety of phosphorus to
the other is not attended with any change of mass, yet it is
accompanied by a change of energy.   When equal masses
of the two kinds of phosphorus are burnt to phosphorus
pentoxide the quantities of heat produced are different.

**Exp. 7.**   Arrange a rough *calorimeter* as shewn in Fig.

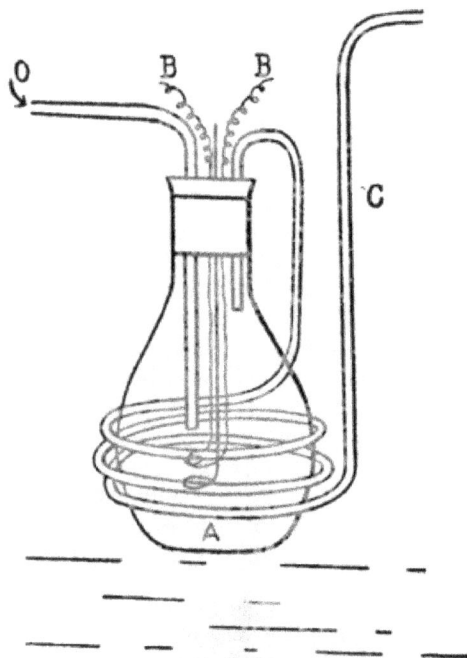

Fig. 48.

48.   *A* is a small, wide-mouthed, short-necked, flask of about
150 c.c. capacity ; it is fitted with a very good ordinary cork,
through which pass, (1) a tube leading to a drying apparatus
which is connected with a gas-holder containing oxygen,
(2) an exit tube (*C*) bent downwards as close to the flask
as possible, and then 2 or 3 times round and close to the
flask, and finally upwards as shewn in the figure, (3) a short
iron spoon round which are wound two silk-covered wires (*B*)
coming from a small electric battery.   These wires terminate
above the cup of the spoon at a distance of about 5 or 8 mm.
apart ; the ends are joined by a little piece of thin platinum,
or iron, wire wound into a little spiral which is then flattened
out to form a support for the phosphorus to be burnt.   Care
must be taken that this thin wire does not touch the iron spoon.
The whole apparatus is immersed in about 300—400 c.c. of
water in a beaker, so that the greater part of the flask is
beneath the surface of the water.   The beaker is placed in
a tin box packed with cotton wool.   It is advisable to place
a shield of thin metal over the end of the cork which is
inside the flask.

A measured, and therefore weighed, quantity of *water*,
which has been for some time in the room where the Exp.
is to be conducted, is placed in the beaker, and in it are
immersed two good thermometers ; a little piece of *phosphorus*
—about ·3 to ·5 gram—is weighed in a very small tube and
is then placed on the support formed by the flattened spiral
of thin platinum or iron wire ; the cork with its tubes &c. in
position is then firmly fixed in the flask ; the flask is lowered
into the water and there clamped ; connexion is made with the
oxygen supply ; the thermometers are read ; a very slow
stream of *oxygen* is passed through the apparatus, and at the
same time the wires (*B*) are connected for a moment with the
battery ; the wire on which the phosphorus is placed becomes
red hot, and the combustion of the phosphorus begins.   By
regulating the rate of the oxygen stream, almost the whole of
the phosphoric oxide produced may be retained in the spiral-
shaped exit tube.   The water is stirred by means of the
thermometers which are read off from time to time ; the temperature attained when the thermometers cease to rise is taken
as the final reading.

As the weight of phosphorus, the weight of water, and the
rise of temperature, are now known, it is easy to calculate the
number of gram-units of heat which have been produced in

the combustion of a specified mass, say 1 gram, of phosphorus.

The apparatus is now cleaned and dried; a fresh piece of thin platinum or iron wire is arranged as a support for the phosphorus; and the experiment is repeated with a weighed quantity of *red phosphorus*; the Exp. being so arranged that the time during which the combustion proceeds is as nearly as may be the same as that occupied in the combustion of the ordinary phosphorus.

This apparatus is not sufficiently accurate to give anything except comparative results; but the results are sufficient to establish the fact that the quantity of heat produced when a specified mass of ordinary phosphorus is completely burnt is considerably greater than that produced when an equal mass of red phosphorus is wholly burnt. The results of Exp. 6 proved that in each of these combustions the same mass of the same compound is produced.

Very many carbon compounds exist each in several forms. A few inorganic compounds also exhibit phenomena similar to those connoted by the term allotropy. The following Exp. illustrates the allotropy of inorganic compounds.

**Exp. 8.** Prepare *mercuric oxide* by heating *mercuric nitrate*, mixed with a little *mercury*, in a basin on a sand-tray so long as acid-fumes are evolved. *This must be done in a good draught place, because of the poisonous effects of vapours containing mercury compounds.*

Also prepare *mercuric oxide* by adding a solution of *mercuric nitrate* to a solution of *potash*, collecting, and thoroughly washing, the pp. which forms.

The oxide prepared by heating the nitrate is a red solid; the other specimen is yellow. Examine the reaction of each with a solution of *oxalic acid*; the yellow oxide dissolves without heating, the red oxide slowly dissolves only on continued boiling. Heat each oxide with an alcoholic solution of *mercuric chloride*; in the case of the yellow oxide a black oxychloride ($HgCl_2 . 2HgO$) is at once produced; in the case of the red oxide no change occurs until after continued boiling.

Now prove that both specimens have the same composition, by determining the mercury in each by the electrolytic method (for details consult a manual of analysis). It is assumed that both compounds are oxides of mercury.

**Exp. 9.** An instructive experiment in allotropic change may be performed by partially converting a measured volume of *oxygen* into *ozone* and measuring the change of volume which attends the change from one form of the element to the other.

Make a small apparatus of glass of about the size shewn in Fig. 49. The apparatus consists of a reservoir of about 5 c.c. capacity with a capillary tube bent as shewn in the figure, and having two platinum wires passing through the glass, as shewn.

Narrow the tube, but do not close it, at the end *a*. Wash the tube thoroughly with nitric acid, then with potash, and finally with distilled water; rinse it out with a little alcohol, and dry it.

Arrange the tube as shewn in Fig. 50, and connect with *a* a small apparatus for evolving pure and dry *oxygen*. Make the oxygen by heating a few grams of pure potassium chlorate in a small tube connected with two U tubes, one containing pumice soaked in potash, and the other pumice soaked in concentrated sulphuric acid. Connect the end *b* (Fig. 50) with a Sprengel's pump. Pass oxygen through the

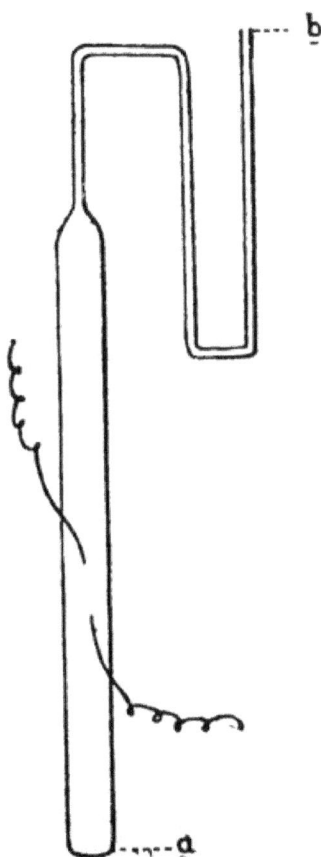

Fig. 49.

apparatus for a little time; then exhaust by the pump and allow the oxygen to pass again; repeat this operation once or twice. Then remove the end *b* (Fig. 50) from connexion with the pump, but maintain the stream of oxygen; bend the end *b* slightly downwards and let it dip into a little concentrated sulphuric acid (oxygen passing through the apparatus all the time); then stop the oxygen-stream; when the acid has risen to about *c* (Fig. 50) seal the tube at *d* and place it upright.

Now place the tube in a bath of water arranging it so that the capillary part hangs over the edge of the bath; after

a time read off the level of the acid in the capillary tube
by bringing a millimetre scale alongside the tube. Seal the
end *b*; pass a silent discharge of electricity through the

Fig. 50.

oxygen in the reservoir as long as the change of level of the
acid in the capillary tube shews that the volume of the oxygen
is diminishing. Then replace the tube in the bath of water;
open the end *b*; and after a time read off the level of the acid,
noting the diminution of volume in the reservoir as so many
mm. on the capillary tube. Re-seal the tube at *b*, and place it
in an air-bath at 300° for 30 mins. or so; then replace the
tube in the water-bath; open *b*, and read off the level of the
acid.

Assuming that oxygen is partially changed to ozone by the
action of the silent electric discharge, this experiment proves that
the change is attended with a diminution of volume, and that
the original volume is almost restored when the ozonised
oxygen is heated to about 300°.

Consult the paper on *Ozone* by Andrews and Tait in
*Phil. Trans.* for 1860, p. 113.

The following numbers shew some of the results there obtained:
(1) 5 c.c. oxygen; diminution = 39·5 mm.; increase on heating to 300° =
38·7 mm. (2) 5 c.c. oxygen; diminution = 37·6 mm.; increase on heating
= 29·7 mm.

*Reference to* "ELEMENTARY CHEMISTRY." Chap. VIII. pars.
118 to 121; also Chap. XVI. pars. 323, 324.

# CHAPTER VII.

## CHEMICAL CLASSIFICATION.

ELEMENTS are placed in the same class because they form compounds having similar compositions and similar properties.

In this chapter are given merely the outlines of some experiments on classification.

I. The three metals Calcium, Strontium, and Barium are placed in the same class.

**Exp. 1.** *Given the oxides,* CaO, SrO, *and* BaO ; *examine quantitatively the interaction of each with water.* To a weighed quantity, 2 or 3 grams, of *each oxide* add *water* drop by drop as long as there is any apparent reaction. Dry the products at about 150° and weigh. Assuming the oxides to have the composition MO (M = Ca, Sr, Ba), and the atomic weights of M and O to be known, calculate the composition of the product of the interaction in each case.

**Exp. 2.** *Examine quantitatively the action of heat on the three hydroxides* $MO_2H_2$. Heat weighed quantities (about ·5 gram in each case) of the compounds prepared by combining the oxides with water, in weighed small hard glass tubes open at one end, to about 200°, 300°, and 400°, weighing the tubes with their contents after each experiment. Also heat weighed quantities of the compounds to about the same temperature in a stream of dry air, freed from $CO_2$, and weigh the products.

**Exp. 3.** *Examine quantitatively the action of heat on the carbonates* $MCO_3$. Heat weighed quantities (about ·5 gram) of the three *carbonates* to redness over a blowpipe-flame for 10 minutes, in open crucibles, and determine the amounts of unchanged carbonates, by dissolving out the oxides produced, by repeated washing with cold water (until the washings do not affect the colour of turmeric paper), drying at 100°, and weighing.

**Exp. 4.** *Prepare the sulphates,* $MSO_4$, *and determine their solubilities in water.* Add the *oxides* to moderately dilute warm *sulphuric acid* until the acid is nearly saturated; remove the remaining acid by washing with cold water, and dry at 150°. Digest an excess of each sulphate in water at the temperature of the air for some hours; then withdraw a measured volume of each solution, evaporate to dryness in water-baths, dry at 150°, and weigh.

**Exp. 5.** *Examine quantitatively the interactions between the sulphates* $MSO_4$ *and saturated solutions of sodium carbonate.* To weighed equal quantities of the *sulphates* add equal volumes of a saturated solution of *sodium carbonate*; boil for equal times; wash the residual mixtures of sulphates and carbonates with cold water as long as the washings contain sulphates; dry at 150°; weigh; determine the carbon dioxide obtainable from each solid, and thus estimate the masses of sulphate and carbonate in each. State your results so as to shew the percentage of each sulphate $MSO_4$ changed to carbonate $MCO_3$.

**Exp. 6.** *Determine, roughly, the heats of neutralisation of aqueous solutions of the hydroxides,* $MO_2H_2$; *the acid being hydrochloric.* Prepare a considerable quantity of a *normal hydrochloric acid solution,* i.e. a solution containing exactly 36·5 grams HCl per litre. Prepare cold saturated solutions of the three hydroxides; keep these solutions in bottles fitted with greased stoppers, each bottle being filled with the liquid; determine the quantity of hydroxide $MO_2H_2$ in a specified volume of each solution, by titration with *standardised oxalic acid solution,* using turmeric paper as an indicator. Calculate the volume of each solution which contains that mass of the hydroxide which is equivalent to the mass of HCl in a specified volume, about 250 c.c. of the normal solution, assuming the interaction to be $MO_2H_2Aq + 2HClAq = MCl_2Aq + 2H_2O$. Measure out this volume of the normal hydrochloric acid solution into each of three beakers; measure out the solutions of the three hydroxides into stoppered bottles of a size such that each is as nearly as may be filled with the solution; place these bottles and the beakers in a box lined and covered with felt 30 to 40 mm. thick, and place this box alongside the *calorimeter* described below, in a room the temperature of which varies but little. After an hour or two, pour the contents of one of the bottles into the calorimeter; quickly add the hydrochloric acid from one of the beakers; stir the liquid, observing the course of the thermometer, and note the highest temperature reached. Wash out and dry the beaker of the calorimeter, and repeat the Exp. with each of the hydroxides.

From your results find the quantity of heat produced when an equivalent, in grams, of each hydroxide ($MO_2H_2$ grams) interacts with an equivalent, in grams ($2HCl$ grams) of hydrochloric acid. In the calculation the *water-equivalent** of the beaker may be taken as the product of the weight of the beaker multiplied by the specific heat of hard glass ($\cdot19$); the solution in the beaker may be assumed to have the same specific heat as water.

The quantity of heat is about 27,500 gram-units in each case.

Calorimeter. Line the inside, bottom, and cover, of a cylindrical pasteboard box with felt about 30 to 40 mm. thick. The box must be of such a size that when a beaker of about 600—700 c.c. capacity is placed in it a space of about 15 to 20 mm. remains between the beaker and the felt. The lid of the box is pierced by two holes; one to admit a thermometer graduated to $\frac{1}{10}$ degrees, the other to admit a stirrer formed of a glass rod reaching to the bottom of the beaker and then bent into a spiral. Weigh a thin beaker of about 650 c.c. capacity; place the beaker in the box, and pack cotton wool round it.

II. The elements Sulphur and Chromium belong to the same group, although they exhibit considerable differences:

**Exp. 7.** *Prepare the trioxides* $SO_3$ *and* $CrO_3$, *and prove that both are acidic oxides.* Arrange an apparatus as shewn in Fig. 51. The bottle *A* contains a concentrated *aqueous solution of sulphur dioxide*; *B* contains concentrated sulphuric

---

* Or the *water-equivalent* may be more accurately determined by adding a known weight of hot water at a known temperature to a known weight of cold water at a known temperature in the calorimeter, and noting the highest temperature which the mixture attains.

acid ; $C$ is a piece of hard glass tubing with some *platinised asbestos* loosely packed into it ; $D$ is a small *dry* flask. The

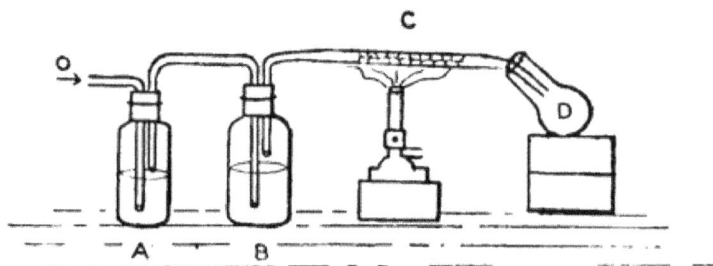

Fig. 51.

asbestos is covered with finely divided platinum by soaking it in concentrated platinic chloride solution and then heating strongly. A stream of *oxygen* is passed into $A$ ; the gas issuing from $A$ is a mixture of oxygen and sulphur dioxide ; the mixed gases are dried by bubbling through $B$, and are then passed over the heated finely divided platinum whereby chemical union occurs ; the sulphur trioxide, $SO_3$, thus produced is condensed in $D$. When a few grams of the white snow-like compound $SO_3$ have collected in $D$, stop the process, and at once dissolve the contents of $D$ in about 500 c.c. water ; *exactly neutralise* a portion of the solution by *potash*; evaporate over a low flame and allow to crystallise ; purify the salt which separates by re-crystallisation from warm water ; dry the crystals and label them p o t a s s i u m s u l p h a t e.

Dissolve 30 grams of commercial *potassium dichromate*, $K_2Cr_2O_7$, in a warm mixture of 50 c.c. *water* and 42 c.c. concentrated *sulphuric acid* ; allow to stand for some hours ; pour off from the pp. of potassium-hydrogen sulphate, $KHSO_4$ ; warm to 80°—90° ; add 15 c.c. concentrated sulphuric acid, and then water drop by drop until the pp. of chromium trioxide which has formed is just dissolved ; then evaporate the liquid to the crystallising point and allow to cool ; after some hours pour off the mother liquor, collect the crystals in a funnel fitted with a small platinum cone pierced with very small holes, and drain by means of the pump ; then spread the crystals of $CrO_3$ on a dry porous tile covered with a bell-jar ; when the crystals are dry remove them by means of a platinum or glass spatula to a beaker, add 5 c.c. of concentrated nitric acid, and spread out the crystals on another porous tile where they are allowed to remain until dry*. Finally warm the crystals to $60^{\circ}$—$80^{\circ}$ until nitric acid ceases to be evolved. Dissolve the chromium trioxide thus prepared in water, neutralise a portion of the solution by *potash*, evaporate, and crystallise. Purify the salt thus obtained by re-crystallisation from warm water; dry the crystals and label them potassium chromate.

Make estimations of potassium and of sulphuric acid in the crystals of potassium sulphate, and estimations of potassium and chromic acid in the crystals of potassium chromate you have prepared. Assuming the atomic weights of potassium, oxygen, sulphur, and chromium, to be known, and assuming the various reactions on which the analytical methods are based to be known, your analyses prove that the compositions of the two salts are represented by the formulæ $K_2SO_4$ and $K_2CrO_4$, respectively.

**Exp. 8.**   Dissolve in water portions of each salt prepared in Exp. 7; to the *sulphate* add an aqueous solution of the *sulphur trioxide*, and to the *chromate* an aqueous solution of the *chromium trioxide*, you prepared; evaporate and crystallise; recrystallise the salts from water; dry, and analyse each. Find the simplest formulæ which express the compositions of the salts.

The salts are $KHSO_4$, and $K_2Cr_2O_7$, respectively.

**Exp. 9.**   *Prepare chromium sulphate from chromium trioxide.* Heat *chromium trioxide* with fairly concentrated *hydrochloric acid* and a little *alcohol*, until the solution is green; allow to cool; add a slight excess of *ammonia*; collect and wash the pale greenish blue pp. thus produced. We shall assume that the composition of this pp. is $Cr_2O_3 . xH_2O$. Dry the pp. at $100^{\circ}$ and dissolve it in excess of *concentrated sulphuric acid*; evaporate until fumes of sulphuric acid begin to be evolved; then allow to cool. Collect the pink solid which separates in a funnel fitted with a little platinum cone pierced with very small holes; dry by spreading on a porous tile in an exsiccator; then wash with alcohol and dry again.

Assuming that the salt thus formed is an anhydrous sulphate of chromium, determine its composition by heating

---

* The object of thus washing with nitric acid is to remove sulphuric acid and $KHSO_4$.

a weighed quantity to full redness so long as it loses weight, weighing the residue, and noting it as $Cr_2O_3$.

The pink salt is $Cr_2(SO_4)_3$.

**Exp. 10.**  *Compare the reaction between sulphuric acid and chromium trioxide with the reaction between the same acid and sulphur trioxide.*

Dissolve about 10 grams of *chromium trioxide* in $1\frac{1}{2}$ parts by weight of *concentrated sulphuric acid* and 3 parts of *water*; place the liquid in a beaker surrounded by cold water, and add *alcohol* drop by drop until the liquid is green; then add a considerable quantity of alcohol; collect the crystals which form and press them between filter paper, and prove, as far as can be done by qualitative reactions, that the salt is a sulphate of chromium.

Now dissolve *sulphur trioxide* in *concentrated sulphuric acid* until the liquid fumes strongly in the air, and set the liquid in an exsiccator over concentrated sulphuric acid. After some time colourless crystals separate; drain these from the liquor; examine their reactions qualitatively, and prove that the substance thus obtained is an acid.

The acid is disulphuric $H_2S_2O_7$.

The results of Exps. 7 to 10 shew that $SO_3$ and $CrO_3$ are acidic oxides; that $CrO_3$ also interacts with acids to form salts the compositions of which shew that they correspond to the oxide $Cr_2O_3$; but that $SO_3$ does not exhibit any basic functions.

**Exp. 11.**  *Prepare chromic chloride and compare it with sulphur chloride.* Dissolve a quantity of moist $Cr_2O_3 . xH_2O$ (*s. Exp.* 9) in hot *concentrated hydrochloric acid*; evaporate until the liquid is thick, and then cool; press the solid which separates between paper; transfer it to a tube of hard glass and heat to about $250^{\circ}$ in a stream of dry chlorine until a lilac-coloured, hard, solid remains. Now raise the temperature to full redness, maintaining a rapid stream of chlorine; let the end of the tube pass into a small weighed bottle; continue to heat strongly until some of the chromic chloride formed is sublimed into the bottle; remove the bottle and at once close it with a good stopper.

The compound thus formed is chromic chloride, $Cr_2Cl_6$.

Compare the composition as determined by your analyses,

the method of formation, and the reactions towards water and acids, of chromic chloride with those of sulphur chloride (s. *Exp. 5, Chap. I. Part II.*).

III.   The elements Iron and Copper belong to different groups.   The following experiments are designed to illustrate some of the differences between these metals.

**Exp. 12.**   *Prepare oxides of the two metals, and determine their compositions.*   Strongly heat a few grams of *finely divided copper* in a stream of *oxygen* so long as any visible change occurs; then heat the product for some time in the air, and allow to cool in an exsiccator.

Perform a similar experiment with very *finely divided iron.*

Estimate the copper in a weighed portion of the copper oxide, either (1) by dissolving in hydrochloric acid, precipitating $CuO_2H_2$ by addition of potash, boiling, washing, heating, and weighing as $CuO$; or (2) by dissolving in nitric acid, adding ammonia, and titrating with a *standardised solution of potassium cyanide.*

Assuming the sole product of the reaction of oxygen with hot copper to be an oxide of copper, find the simplest formula which expresses the composition of this oxide ($Cu = 63.2$, $O = 16$).

Estimate the iron in a weighed portion of the iron oxide prepared, by dissolving in hydrochloric acid, reducing the ferric chloride so formed to ferrous chloride by means of *stannous chloride*, and titrating with a *standardised solution of potassium dichromate.*   Assuming the sole product of the interaction of iron and oxygen at a high temperature to be an oxide of iron, find the simplest formula which expresses the composition of this oxide ($Fe = 56$, $O = 16$).

**Exp. 13.**   *Determine the temperatures at which reduction of the oxides* $CuO$ *and* $Fe_2O_3$ *by CO begins.*

[*Before beginning this Exp. read the paper by Wright and Luff in* C. S. JOURNAL, TRANS. for 1878, p. 1.]

Prepare 10 or 15 grams of $CuO$, and about the same quantity of $Fe_2O_3$, by adding pure potash to $CuSO_4Aq$ and $Fe_2Cl_6Aq$, respectively, boiling, washing the pps. thoroughly by hot water, drying and heating over a Bunsen-lamp.   Place about ·5 gram of the *copper oxide* and the *ferric oxide* thus prepared in test tubes connected with potash bulbs containing clear, freshly prepared, *baryta water*; immerse the tubes in a paraffin bath; pass pure *carbon monoxide* over each oxide, and determine, in each case, (1) the temperature at which a faint turbidity is produced in the baryta water, i.e. the temperature at which the reduction of the metallic oxide by carbon monoxide begins, and (2) the temperature at which a considerable turbidity is produced by a few bubbles of the gas issuing from the tubes, i.e. the temperature at which the reducing action of the carbon monoxide is well marked.

*For details of the method of procedure reference must be made to the paper referred to above.*

The results of this Exp. shew that cupric oxide is reduced by carbon monoxide at a temperature considerably lower than that at which ferric oxide is reduced.

**Exp. 14.**   *Examine the reactions between steam and the metals copper and iron.*

Place about 10 grams of the oxides CuO and $Fe_2O_3$, prepared in the last Exp., in tubes of hard glass, each of which is connected with an apparatus for preparing *pure dry hydrogen.* The hydrogen should be prepared by the reaction of pure dilute sulphuric acid with redistilled zinc, and should be purified from any hydrides of sulphur, carbon, phosphorus, and arsenic, by passage through solutions of lead acetate, and copper sulphate made strongly alkaline by potash.   Heat the oxides in the hydrogen-stream until reduction to metal is complete.   Then remove the hydrogen apparatus; replace it by a flask containing hot water; fit corks and exit tubes into the other ends of the tubes containing the metals; arrange the tubes so that each passes through a small iron box fitted with a thermometer; pass *steam* over each metal, and roughly compare the temperatures at which the steam is decomposed, with production of hydrogen, by the two metals.

Iron decomposes steam fairly rapidly at a low red heat; copper, only at a full red heat, and then slowly.

**Exp. 15.**   *Prepare sulphates of the two metals, determine their compositions, and compare the reaction of each with an oxidising agent.*

Dissolve about 20—30 grams of *cupric oxide* in *warm dilute sulphuric acid*; evaporate to the crystallising point; pour off the mother liquor from the crystals which form on cooling;

recrystallise two or three times from water, and dry by pressing between filter paper.

Estimate the copper in a weighed quantity of these crystals. Also estimate the water by heating to $210^\circ$—$220^\circ$ and weighing the dehydrated copper sulphate which remains. Before estimating the water prove that the crystals do not evolve sulphuric acid when heated over a Bunsen-lamp.

Prepare crystals of ferrous sulphate by adding *iron filings* little by little to *warm dilute sulphuric acid* until no more iron is dissolved, then adding a little more iron, boiling for a few minutes, filtering into a beaker which has been rinsed out with a little concentrated sulphuric acid (this dissolves any basic sulphate that is formed during evaporation), adding *alcohol* to the warm liquid, stirring well, collecting the pp. on a filter, washing with alcohol, spreading out on filter paper, and exposing to the air of a moderately warm room until the smell of alcohol is quite gone. Preserve the crystals in a well stoppered bottle.

Estimate iron in a weighed quantity of the crystals by dissolving in water, adding sulphuric acid, and *at once* titrating with *standardised permanganate solution.* Heat some of the crystals over a Bunsen-lamp and prove that sulphuric acid is evolved. The water cannot therefore be estimated by determining the loss of weight which the crystals undergo when heated. Mix a weighed quantity of the crystals with 2 or 3 times their weight of dry pure *lead monoxide*; place the mixture in a short tube of hard glass, and cover it with a layer of lead monoxide; weigh the tube and its contents; heat to about $300^\circ$ until moisture is no longer given off; cool in an exsiccator, and weigh again. Repeat the heating until the weight of the tube is constant. (The lead oxide combines with the sulphuric acid produced by heating, and forms lead sulphate—$PbSO_4$,—which is unchanged at the temperature of the experiment.)

Find the simplest formulae which express the compositions of the two sulphates you have prepared (Fe = 56, Cu = 63·2, O = 16, S = 32.)

The sulphates are $CuSO_4.5H_2O$, and $FeSO_4.7H_2O$, respectively.

Now roughly weigh out a few grams of the *copper sulphate* you have prepared; dissolve in water; add about ⅕ as much *concentrated sulphuric acid* as the weight of the copper sulphate taken; boil, and add *concentrated nitric acid* drop by drop; evaporate, and crystallise; recrystallise from water; dry the crystals obtained, and determine copper in them. These crystals contain the same percentage of copper as those with which you began the Exp.; their appearance and prominent physical properties are also the same as those of the original crystals.

Copper sulphate $CuSO_4.5H_2O$ is unchanged by the action of oxidising agents.

To a solution in water of 10 or 15 grams of the *ferrous sulphate* you have prepared add about ⅓ the weight of the salt of *concentrated sulphuric acid*; then heat to boiling, and add *concentrated nitric acid* drop by drop until the colour of the liquid is a clear reddish yellow; now evaporate until the liquid gets thickish; then allow to cool, and pour off the liquid from the semi-solid mass which remains; heat the residue over a *very low* Bunsen-flame until it is quite dry, and sulphuric acid is no longer evolved.

The appearance, and comparative insolubility in water, of the solid thus produced shew that it is not ferrous sulphate. Dissolve an accurately weighed quantity of this solid (about ·5 gram) in dilute sulphuric acid; add a few pieces of *pure zinc*, and allow the reaction to proceed until the liquid is almost colourless, and all zinc is dissolved; then determine the iron by titration with *permanganate*. Assuming the salt to be anhydrous, and to be a sulphate of iron, calculate its composition, and find the simplest formula which will express this composition.

The salt prepared by the interaction of nitric and sulphuric acids with ferrous sulphate is ferric sulphate, $Fe_2(SO_4)_3$: the nitric acid supplies oxygen;

$$2(FeO.SO_3) + O + SO_3.H_2O = Fe_2O_3.3SO_3 + H_2O.$$

Before beginning Part III. the student should work out a number of fairly difficult quantitative analyses; he should also go through a systematic course of preparations of organic compounds; and a course of gas-analysis. [He may also, if time permit, perform a few qualitative analyses of complex mixtures of salts including the salts of the *rare* elements.]

While the student is performing the experiments described in Part III. he should also be working at practical physics; in his work in the physical laboratory he should pay particular attention to determinations of (1) refractive indices of organic compounds, (2) specific rotatory powers of organic compounds, (3) wave-lengths and also mapping spectra, (4) quantities of heat, (5) electrical conductivities of solutions of acids and salts, (6) electrolytic depositions of different metals, (7) crystallographic forms.

# PART III.

## CHAPTER I.

**Exp. 1.** Determination of the atomic weights of copper and iron.

Prepare and analyse *the two oxides of copper*, and *the two chlorides of iron*, and thus find the combining weights of copper and iron, i.e. the masses which combine with 8 parts by weight of oxygen, or 35·5 parts by weight of chlorine.

Purify *copper sulphate* by solution in water, adding a little *nitric acid*, and heating to oxidise ferrous salts present, digesting with cupric oxide to precipitate ferric oxide, filtering, and crystallising. Recrystallise the crystals thus obtained, dissolve in water, heat to boiling, and add *pure potash solution* drop by drop until only a trace of copper sulphate remains undecomposed. Wash the pp. thoroughly by decantation with boiling water, dry at 100°, then heat over a Bunsen-lamp, and preserve the cupric oxide in a stoppered bottle.

Dissolve 20 grams pure *copper sulphate*, 30 grams *cream of tartar*, and 40 grams *dextrose*, in a basin, in 240 c.c. water; heat, and add 30 grams *caustic soda*, and boil until the blue liquid is colourless; wash the red pp. of cuprous oxide thoroughly by decantation with boiling water, dry at 100°, and preserve in a stoppered bottle.

Prepare ferrous chloride by dissolving *very thin piano wire* in *pure hydrochloric acid*, evaporating in contact with iron till the liquid begins to get thick, filtering, evaporating the filtrate rapidly to dryness in a basin, mixing the solid with about an equal quantity of *pure ammonium chloride*, filling a flask of about 150 c.c. with the mixture, and heating over a large

11—2

lamp until the ammonium chloride is completely volatilised ; on cooling the semi-fused mass, crystals of ferrous chloride are formed ; place the crystals in a stoppered bottle.

Prepare ferric chloride by strongly heating *very thin iron wire* in a rapid stream of *dry chlorine* ; sublime the crystals, in a stream of chlorine, into a dry stoppered bottle.

Estimate copper and oxygen in the *oxides of copper* by heating about ·5 gram of each, placed in a porcelain boat in a tube of hard glass, in a stream of *pure dry hydrogen* (*s. Exp.* 2 *in Chap. III. of Part II.*). Weigh the residual copper in each case, and determine the oxygen by loss. Calculate the weight of copper combined with 8 parts by weight of oxygen in each oxide. Estimate iron in the *chlorides of iron* by dissolving about ·5 gram of each in very dilute hydrochloric acid, and titrating with a *carefully standardised permanganate solution* : reduce the ferric chloride to ferrous, by means of *pure zinc*, before titration with permanganate. Also estimate the chlorine in the chlorides of iron by dissolving about ·5 gram of each in *dilute nitric acid*, adding a slight excess of *pure silver nitrate solution*, and determining the mass of silver chloride precipitated by the ordinary gravimetric method. Calculate the weight of iron combined with 35·37 parts by weight of chlorine in each chloride.

The results of your experiments shew that 31·6 and 63·2 parts by weight of copper combine with 8 parts by weight of oxygen to form cuprous and cupric oxide, respectively; and that 28 and 18·6 parts by weight of iron combine with 35·37 parts by weight of chlorine to form ferrous and ferric chloride respectively. The atomic weight of copper is then $x$31·6, and the atomic weight of iron is $n$18·6 and $m$28, where $x$, $n$, and $m$ are whole numbers.

Measurements of the specific heats of the two metals will serve to determine the values of $x$, $n$, and $m$.

Prepare approximately *pure iron* as follows. Place 100 or 200 grams of thin iron wire in small pieces in a Hessian crucible and heat in the air until a portion of the iron is superficially oxidised ; then cover the contents of the crucible with a plate of green glass, and heat to the highest temperature of a good wind-furnace. The oxygen of the ferric oxide combines with the small quantities of carbon in the iron, and the excess of ferric oxide dissolves in the molten glass.

Prepare approximately *pure copper* by electrolysing a solution of pure copper sulphate acidulated with sulphuric acid, using platinum poles; wash the copper which separates on the negative electrode and hammer it in a clean steel mortar into a compact mass.

Arrange a *calorimeter* as described in Chap. VII. Part II.; place about 20—30 grams of the iron you have prepared in a thin wide test tube; close the mouth of the tube by a plug of cotton wool through which passes a thermometer with its bulb in contact with the metal; immerse the bulb in steam until the temperature of the metal is constant; note the temperature of the metal and of the water in the calorimeter; quickly transfer the metal to the calorimeter, and determine the rise of temperature. Repeat a similar experiment with about 20—30 grams of the copper you have prepared.

Calculate the specific heat of each metal, taking into account the water equivalent of the calorimeter stirrer and thermometer used.

Use the results thus obtained to determine what multiples of the combining weights of copper and iron, respectively, represent the atomic weights of these elements.

### Exp. 2. Determination of the atomic weight of silver, that of chlorine being known; or of the atomic weight of chlorine, that of silver being known.

This determination resolves itself into finding the ratio of the masses of silver and chlorine which interact to form silver chloride, when silver nitrate solution is added to a solution of ammonium chloride. Read STAS, *Recherches sur les rapports réciproques des poids atomiques*, p. 38 *et seq*; or the German translation *Untersuchungen über die Gesetze der chemischen Proportionen* &c., pp. 31—60.

It is necessary to prepare *pure silver*, *pure ammonium chloride*, and *pure sodium chloride*; the measuring vessels used must be carefully corrected.

Pure Silver. Dissolve ordinary *coinage silver* (containing copper) in dilute, boiling, nitric acid; evaporate to dryness, and heat till the mass melts; after cooling, dissolve in water containing ammonia, and allow to stand for 48 hours; then filter through a double filter, and dilute until the liquid does not contain more than 2 p.c. of silver.

Measure the total volume of the liquid.

Prepare a solution of ammonium sulphite by saturating pure ammonia solution with sulphurous oxide. The *sulphurous oxide* is prepared by heating *pure sulphuric acid*, diluted with

$\frac{1}{2}$ to $\frac{2}{5}$ its volume of *water*, with *pure copper*, and leading the gas through water. Add excess of ammonia to a small measured quantity of the ammonium sulphite solution ; heat to boiling, and run in the ammoniacal solution of silver nitrate containing copper nitrate so long as the blue colour of this solution is removed. Cuprous sulphite is formed and this reduces the silver nitrate with precipitation of silver.

Having thus found the volume of the solution of silver and copper nitrates which is decolourised by a specified volume of the ammonium sulphite solution, add to the whole of the solution of silver nitrate containing copper nitrate that volume of the ammonium sulphite solution which will just suffice to decolourise it, place the liquid in a clean stoppered bottle and immerse it in water at 60°—70°. The whole of the silver in the solution is soon precipitated ; wash the pp. by decantation with ammoniacal water until the washings cease to give a pp. with barium chloride solution ; then allow the pp. to remain in contact with concentrated ammonia solution for a few days; finally wash thoroughly with cold water, and dry.

Ammonium chloride. One litre of a boiling concentrated solution of *ammonium chloride* is mixed with 100 c.c. of *nitric acid*, specific gravity 1·4 ; boiling is continued so long as chlorine is evolved, and the liquid is then allowed to cool ; the ammonium chloride which separates is dissolved in boiling water, $\frac{1}{20}$ volume of nitric acid is added, and boiling is continued so long as chlorine is evolved. The liquid is then diluted with so much water that no solid separates out on cooling, run into a retort containing *pure slaked lime*, and distilled. The ammonia thus produced is washed by passing through a little water, and is then led into cold water. This solution is nearly saturated with *hydrochloric acid gas*, made by warming pure hydrochloric acid. The liquid is concentrated ; the ammonium chloride which separates is dried at 100°, in a long necked flask of hard glass, through which is passed a slow stream of pure dry ammonia gas. The dry salt is sublimed at the lowest possible temperature, the neck of the flask being kept full of ammonia gas.

Sodium chloride. Pure *bicarbonate of soda* is heated to dull redness in a platinum dish ; the residue is dissolved in cold water ; the liquid is evaporated in a platinum dish and crystallised; this process is repeated three or four times. The solid is dissolved in cold water, and saturated with *pure hydrochloric acid gas* ; the sodium chloride which separates is dried, heated to dull redness in a platinum dish, and dissolved in water; the solution is allowed to stand for 24 hours, and is then poured off from the silica which has settled down, and evaporated in a platinum vessel; the residue is heated to dull redness and cooled in an exsiccator.

A *decinormal solution* of the sodium chloride is prepared; 5·837 grams in 1000 c.c.: 1 c.c. precipitates ·000766 grams of silver.

Method of procedure. A wooden box about 70 centims. long, 22 centims. wide, and 45 centims. deep, is arranged with doors and openings as shewn in Figs. 52 and 53. The inside

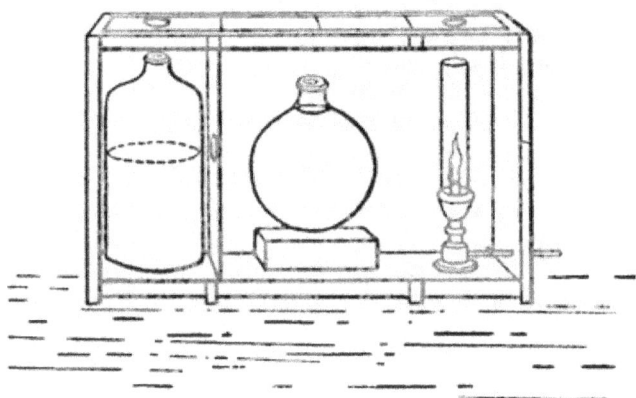

Fig. 52.

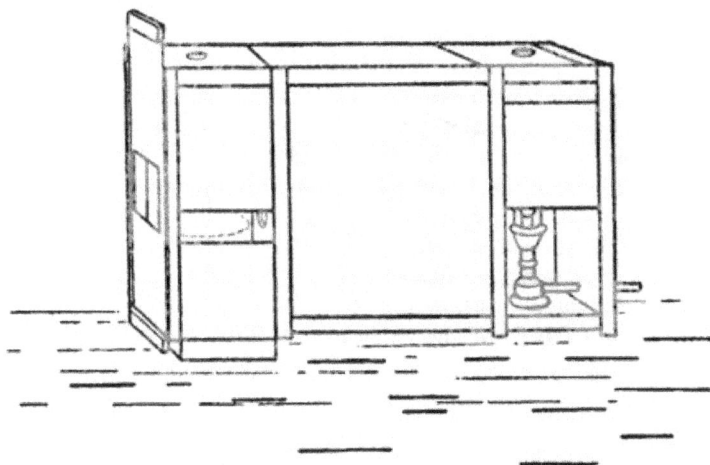

Fig. 53.

of the box is blackened. The flask or bottle containing the
solution of silver to be estimated is placed on one side of, and
close to, the screen which divides the box into two parts, and
at such a height that the surface of the liquid in this flask
is level with the centre of the round opening (3 centims.
diameter) in the screen. A spherical flask containing a solution of *potassium chromate* is placed on the other side of the
screen, and a lamp is placed behind this flask; this arrangement serves to throw a pencil of yellow light through the
opening in the screen into the upper layers of the liquid in the
flask.

Weigh out *very carefully* about 1 gram of *pure silver* into
a flask of about 300 c.c. capacity; fit the flask with a caoutchouc cork previously boiled in caustic soda, and well washed,
carrying a tube with a bulb blown on it and then bent twice
at right angles; place the flask in water kept at 40°—45°, and
allow the open end of the tube to dip into water in a beaker;
now add enough *pure nitric acid*, specific gravity 1·24, to dissolve the silver; when solution is complete, allow the flask to
cool, and water to flow back into the flask; wash the exit
tube once or twice with water. When the contents of the
flask are cold add *pure ammonia*, drop by drop, until the
liquid smells of ammonia, then add *pure dilute acetic acid* until
the smell of ammonia is removed. Place the flask in the '*titra-
tion-box*'. Weigh out *very carefully* a quantity of *ammonium
chloride* rather less than that which the silver in solution
is capable of completely decomposing ($NH_4ClAq + AgNO_3Aq$
$= (NH_4)NO_3Aq + AgCl$; 107·66 Ag decompose 53·38$NH_4Cl$);
add this to the liquid in the flask; shake thoroughly and allow
the silver chloride to settle. There now remains a small quantity of silver in solution; to estimate this, fill a 50 c.c. burette
graduated to $\frac{1}{20}$ c.c. with the decinormal sodium chloride solution; arrange the burette so that the liquid can be dropped
into the flask through the opening in the lid of the box;
light the lamp, and arrange matters so that the upper layer of
liquid in the flask is illuminated by the yellow light; add the
*sodium chloride solution* one drop at a time; shake the flask
well after each drop has run in, and allow to settle; continue
the addition of sodium chloride until a drop of the solution
ust ceases to produce an o palescence in the clear liquid in the
flask. $\frac{1}{20}$ mgm. of silver in 1000 c.c. water may be thus determined with accuracy.

Calculate the weight of silver remaining in solution after

addition of the ammonium chloride; then find the weight of
silver which has been used to precipitate the chlorine of 100
parts by weight of ammonium chloride. From these results
calculate (1) the atomic weight of silver if $Cl = 35\cdot37$ ; (2) the
atomic weight of chlorine if $Ag = 107\cdot66$ ; assuming that the
ratio of the atoms of silver and chlorine in silver chloride is
$1 : 1$.

# CHAPTER II.

## DISSOCIATION.

THE molecular weights of gasifiable elements and compounds are determined by finding the specific gravities of their vapours (*s. Part II. Chap. III. Exp.* 1). But some of the results obtained appear at first sight to be abnormal. The greater number of these apparently abnormal results are explained by the occurrence of dissociation. The phenomena of dissociation were briefly studied in Part II., but some other and more accurate Exps. must now be performed. (*s.* Pattison Muir's *Elements of Thermal Chemistry*, Chap. IV. Sect. 2.)

**Exp. 1. Dissociation of hydriodic acid.** Read Lemoine, *Équilibres chimiques entre l'hydrogène et l'iode gazeux*, Annales de Chim. et Phys. (5). 12, 145.

I. Lead a rapid stream of *hydriodic acid gas* for half an hour or so into a dry bulb (Fig. 54) of about 250 c.c. capacity, blown from the same glass as that from which pressure tubes are made. During the operation of filling, the bulb is surrounded by water of a known temperature. The neck of the bulb has been previously narrowed and thickened at *a*. The entrance of

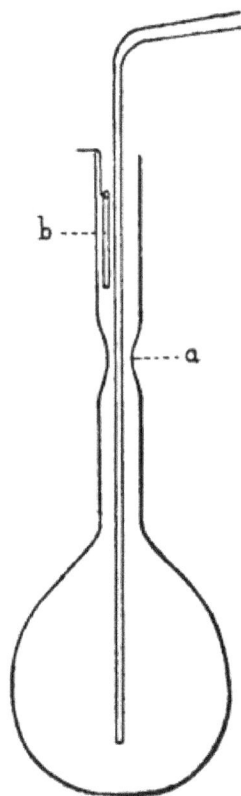

Fig. 54.

moisture during the filling of the bulb with gas is prevented
by suspending in the neck a small tube, $b$, filled with calcium
chloride. The gas is conveniently prepared from a mixture
of 120 grams iodine, 6 grams amorphous phosphorus, and
40 grams of a concentrated solution of hydriodic acid; the
gas must be thoroughly dried by passage over phosphorus
pentoxide. When the bulb may be presumed to be full of the
gas it is sealed at $a$; at the same time the barometric pressure
is read. The bulb is now heated for a couple of hours or more
in the vapour of boiling sulphur ($440^\circ$); and then cooled as
rapidly as possible by wiping it with a damp cloth.

The sealed end of the neck is now broken off under a *saturated solution of common salt* which has been well boiled and
cooled out of contact with the air. The iodine resulting from
the dissociation of the hydriodic acid, as well as any undecomposed hydriodic acid, dissolves in the saline solution;
the residual gas in the bulb is transferred to a graduated tube
and its volume is read off, and reduced to $0^\circ$ and 760 mm. ($= v$).
It is necessary to transfer some of this gas to a eudiometer
and analyse it, as it is impossible to get the bulb perfectly
air-free. The volume of the bulb $v'$ must also be accurately
determined by weighing it full of water. We have now all
the data necessary for determining the percentage dissociation
of hydriodic acid under the conditions of the experiment.

The following example of the mode of calculation is taken from
Lemoine's paper.

$v = 35\cdot48$ c.c. This gas on analysis gave $96\cdot7\,^0/_0$ H and $3\cdot3\,^0/_0$ N.
Hence $v$ consisted of $34\cdot31$ c.c. H and $1\cdot18$ c.c. N (the latter number
corresponds to $1\cdot5$ c.c. of air).

The bulb, which had a capacity $v' = 394\cdot8$ c.c., had been filled with
hydriodic acid gas at a temp. of $10\cdot7^0$ and under a pressure of $748\cdot2$ mm.,
hence it follows that the amount of gas operated on would occupy at $0^0$
and 760 mm.

$$\frac{394\cdot8 \times 748\cdot2}{1\cdot039 \times 760} = 374\cdot1 \text{ c.c.}$$

Now these $374\cdot1$ c.c. must have contained $1\cdot5$ c.c. of air. Hence the
true quantity of III used was $374\cdot1 - 1\cdot5 = 372\cdot6$ c.c. The half of this
$= 186\cdot3$ c.c. represents the hydrogen of this hydriodic acid; but on
warming, the $\cdot32$ c.c. of oxygen in the air present would oxidise $\cdot64$ c.c.
of this hydrogen; hence there are only $185\cdot7$ c.c. of "disposable" hydrogen.
Hence the ratio of free hydrogen to total disposable hydrogen (which is a
measure of the dissociation)

$$= \frac{34\cdot31}{185\cdot7} = \cdot185.$$

At the temperature of 350° at which this experiment was conducted it is easy to see that the pressure must have been $\frac{374 \cdot 1}{394 \cdot 8} \times \frac{623}{273}$ or about 2·15 atmospheres.

Now perform another experiment at 440°; but this time let the hydriodic acid gas be introduced into the experimental bulb at such a diminished pressure as will become about an atmosphere at 440°; this is done by connecting the neck of the bulb with a three-way tube, the other limbs of which are connected, respectively, with a vacuum-pump, and a large globe containing dry hydriodic acid. In this case it will be necessary to leave the bulb in sulphur vapour for 25 hours or so. The warming may be effected in periods of 12 hours each, or less, provided after each warming the bulb and contents are cooled rapidly.

*The results of the two experiments ought to shew that the limit of dissociation of hydrogen iodide at 440° varies very slightly indeed with the pressure of the reacting gases.*

II. Perform two experiments similar to the above, but heat the bulbs to a temperature of about 360°, for say 8 hours. In neither case will equilibrium between the two inverse actions

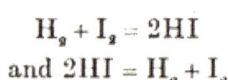

$$H_2 + I_2 = 2HI$$
$$\text{and } 2HI = H_2 + I_2$$

be reached, as was the case at 440°; but your experiments ought to shew *that in the bulb which has been submitted to the lower pressure there has been more dissociation than in that subjected to the higher pressure.*

The constant temperature of 360° may be obtained by immersing the bulbs in the vapour of boiling mercury, contained either in a cylindrical vessel of sheet iron of such a length that the mercury vapour is condensed in its upper cooler portions, or preferably in an iron bottle of the kind designed by Deville for use in vapour density determinations. (*Annales de Chim. et Phys.* (3), 58, 257.)

III. Find as accurately as possible the volumes of two globes. Into one of these put as much *iodine* as is equivalent to the amount of hydrogen the globe will contain under atmospheric pressure, and into the other just half as much iodine as is equivalent to the amount of hydrogen the globe will contain under ordinary pressure [100 c.c. H at 0° and 760 mm. are equivalent to 1·137 grams I]. Fill both globes with *dry hydrogen*, in the manner described under I. for hydriodic acid: seal, and heat to 440° for 4 hours. Then cool quickly, and determine the ratio of free hydrogen to total hydrogen as before.

Now calculate the ratio of hydriodic acid formed to hydriodic acid possible, in each case ; from the equation

$$\frac{\text{hydriodic acid formed}}{\text{hydriodic acid possible}} = \frac{1 - \lambda}{\omega},$$

where

$$\lambda = \frac{\text{hydrogen free}}{\text{hydrogen total}} \text{ and } \omega = \frac{\text{number of equivs. of iodine}}{\text{number of equivs. of hydrogen}}.$$

Your experiments ought to shew that as ω decreases, λ increases ; or, in other words, that, *other conditions being the same the dissociation-limit is lessened when either of the products of dissociation is present in excess.*

## Exp. 2.  Dissociation of ammonium carbamate.  Read Horstmann, *Annalen der Chem. u. Pharm.* 187, 48.

I. Dissociation of ammonium carbamate in vacuo.

Procure a tube, closed at one end, about a metre in length and about 15 mm. in diameter. The tube is provided with a millimetre scale in terms of which it is accurately calibrated. Having thoroughly cleaned and dried this tube, lead into it perfectly dry *ammonia* and *carbon dioxide* gases. Make the delivery tubes so long that the gases first come into contact with each other near to the closed end of the tube. White crystalline plates of ammonium carbamate $CO(NH_2)(ONH_4)$ form on the sides of the experimental tube. When this formation has gone on for some time, fill the tube very carefully with perfectly dry (but not warm) mercury, and invert it in a trough of mercury. It will be found that the height of the mercury in the experimental tube is now less than the height of the barometer. This is due to the fact that at ordinary temperatures, i.e. 17°—20°, the carbamate undergoes a notable amount of dissociation into carbon dioxide and ammonia.

The difference between the barometric height and the height of mercury in the experimental tube gives the dissociation-pressure for the temperature considered.

The following are some of the values obtained by Horst-

mann for the dissociation-pressures of the carbamate at ordinary
summer temperatures ;—

| $t^0$ | mm. |
|-------|------|
| 17·6 | 52·6 |
| 18·5 | 55·7 |
| 20 | 64·7 |
| 20·4 | 67·6 |

It is evident from these numbers that the dissociation
increases with the temperature.

The student should perform three or four experiments
in the manner described to verify this result.

II.   Dissociation of ammonium carbamate in presence of excess of either of its products of dissociation.

It can be theoretically proved (see Horstmann *loc. cit.*)
that although the dissociation-pressure of ammonium carbamate in presence of excess of *either* of its products of dissociation ought to be less than what it would be *in vacuo* at
the same temperature, yet an excess of ammonia ought to
diminish the dissociation-pressure to a greater extent than an
equal excess of carbon dioxide.

Prove this experimentally.   The apparatus required is the
experimental tube already described and a small gasometer
similar to those used in nitrogen-determinations ; the gasometer is calibrated with the same mass of mercury as was used
for the calibration of the experimental tube.   The *perfectly
dried* gas (either *ammonia* or *carbon dioxide* as the case may
be) is brought into the gasometer and its volume is read off.
The end of the delivery tube of the gasometer is then brought
under the mouth of the experimental tube, and gas is forced
into this by pouring mercury into the funnel tube.   When
sufficient gas has passed into the experimental tube the volume
of residual gas in the gasometer is read off.   We have now
all the data necessary for calculating the partial pressure, $P$,
which the gas introduced into the experimental tube would
exert there.

The difference between the total pressure $\pi$ in the tube
when this has become constant and the partial pressure $P$
gives us the dissociation-pressure $(p_1)$ of the carbamate under
the conditions of the experiment.

It is scarcely necessary to add, that the experiments
should be conducted in a room the temperature of which

remains or can be maintained as nearly as possible constant
for intervals of 3 or 4 hours.

Appended are some of the results obtained by Horstmann.

### Excess of $CO_2$.

| $t^0$ | $\pi$ | $P$ | $p_1$ | $p$ |
|---|---|---|---|---|
| 20·4 | 78·9 | 33·5 | 45·4 | 66·2 |
| 21·8 | 105·8 | 69·9 | 35·9 | 71·0 |
| 18·3 | 112·4 | 87·4 | 25·0 | 55·0 |

### Excess of $NH_3$.

| $t^0$ | $\pi$ | $P$ | $p_1$ | $p$ |
|---|---|---|---|---|
| 21·8 | 69·5 | 24·4 | 45·1 | 70·9 |
| 20·6 | 75·2 | 35·4 | 39·8 | 65·3 |
| 17·8 | 359·8 | 355·3 | 4·5 | 53·0 |

In these tables $p$ denotes the dissociation-pressure of the
carbamate *in vacuo*.

Four or five experiments should be made with different
partial pressures in the case of each gas.

If the results in each case are plotted out in such a way
that the values for the ratio $\dfrac{P}{p}$ are measured along ordinates,
and the values for the ratio $\dfrac{p_1}{p}$ along abscissæ, two curves are
obtained which are independent of the various temperatures
at which the experiments have been conducted. In accordance
with what has been said above, it will be found that the
curve for ammonia is steeper than that for carbon dioxide.

# CHAPTER III.

WHEN equivalent masses of two acids and a base are allowed to interact in dilute aqueous solution, the base divides itself between the acids in a definite ratio; this ratio expresses the relative affinities of the acids for the base. The values of the relative affinities of acids quantitatively condition many, if not all, chemical changes which are brought about by the acids. (*s.* Pattison Muir's *Principles of Chemistry*, Book II. Chap. III., where references are also given to original memoirs of importance.)

**Exp. 1. Thermal methods of measuring relative affinities of acids.** Read Thomsen, *Thermochemische Untersuchungen*, 1, 97—126.

When equivalent masses of sulphuric acid and caustic soda interact in dilute aqueous solution, $a$ gram-units of heat are produced; when equivalent masses of nitric acid and soda interact, $b$ gram units of heat are produced. When equivalent masses of the two acids and soda interact, either the whole of the soda forms sulphate, in which case $a$ units of heat are produced; or the whole of it forms nitrate, in which case $b$ units of heat are produced; or both sulphate and nitrate are formed, in which case the quantity of heat produced is different from either $a$ or $b$.

Measurement of the heat actually produced will furnish data from which conclusions may be drawn regarding the distribution of the soda between the two acids.

When equivalent masses of sodium sulphate and nitric acid interact in dilute aqueous solution, either no change occurs, in which case no thermal disturbance is noticed; or the whole of the sulphate is decomposed, in which case a thermal change, equal in amount to the difference between $a$ and $b$, occurs; or a portion of the sulphate is decomposed, in which case the amount of thermal change is less than the whole difference between $a$ and $b$.

Supposing that $b < a$, and that when sodium sulphate and nitric acid interact $x$ grams-units of heat disappear, $x$ being less than $b - a$; then, if no other changes occur than those taken into account, we may say that $\dfrac{x}{b-a}$ of the total mass of sodium sulphate has been decomposed; hence we might calculate the distribution of the soda between the two acids, and so find the relative affinities of the acids. Thus, Thomsen found $a = 31,378$, $b = 27,234$ ($b - a = -4144$), $x = -3504$; hence, assuming that no changes occur except those represented by $a$ and $b$, $\dfrac{-3504}{-4144} = \cdot845$ of the total sodium sulphate has been decomposed by the nitric acid.

But if, when sodium sulphate and nitric acid interact, the whole of the sulphate is not decomposed, the solution must contain sodium sulphate and nitrate, and also sulphuric and nitric acids; either or both of the acids may interact with either or both of the salts, or the acids may interact with each other, and these changes must be accompanied by production or disappearance of heat. Thomsen has found that the only one of the possible reactions accompanied by more than a very small thermal change is that between the sulphuric acid and the sodium sulphate. Hence it is possible to find by thermal methods the approximate distribution of soda between two acids when the three bodies interact in dilute solution in equivalent masses.

The data to be determined are the thermal values of the following chemical changes;— (1) $[H^2SO^4Aq, 2NaOHAq]$; (2) $[2HNO^3Aq, 2NaOHAq]$; (3) $[Na^2SO^4Aq, 2HNO^3Aq]$; (4) $[Na^2SO^4Aq, nH^2SO^4Aq]$.

Thomsen's solution of the problem is given to shew how the calculations are made.

$$(1) = 31,378; \quad (2) = 27,234, \therefore (2) - (1) = -4144;$$

$$(3) = -3504; \quad (4) = -\frac{n}{n + \cdot8}\, 3300.$$

The reaction between equivalent masses of sodium sulphate and nitric acid may be expressed thermally thus

$$[Na^2SO^4Aq, H^2N^2O^6Aq]$$
$$= x([H^2N^2O^6Aq, Na^2O^2H^2Aq] - [H^2SO^4Aq, Na^2O^2H^2Aq])$$
$$+ (1-x)[Na^2SO^4Aq, \frac{x}{1-x}H^2SO^4Aq].$$

Substituting Thomsen's values, we have

$$[Na^2SO^4Aq, H^2N^2O^6Aq]$$
$$= x \times -4144 + (1-x)[Na^2SO^4Aq, \frac{x}{1-x}H^2SO^4Aq].$$

If $x$ is assumed $=\frac{2}{3}$, the calculated value of this equation becomes $-3546$; the observed value of the reaction between $Na_2SO_4Aq$ and $H_2N_2O_6Aq$ was $-3504$; hence $x$ is probably $=\frac{2}{3}$; i.e. the affinity of nitric acid for soda is twice as great as that of sulphuric acid for the same base; or in other words, when equivalent masses of soda, nitric acid, and sulphuric acid, interact in dilute aqueous solution, $\frac{2}{3}$ of the soda forms sodium nitrate, and $\frac{1}{3}$ forms sodium sulphate.

Method of procedure. Solutions of *caustic soda, sulphuric acid, nitric acid*, and *sodium sulphate*, are required; convenient strengths are (1) 40 grams NaOH in 1800 grams water, (2) 98 grams $H_2SO_4$ in 3600 grams water, (3) 63 grams $HNO_3$ in 1800 grams water, (4) 142 grams $Na_2SO_4$ in 3600 grams water.

The soda solution may be prepared by dissolving rather more than the proper quantity of '*soda purified by alcohol*' in water, adding *baryta water* drop by drop so long as a pp. of barium carbonate is produced, then a very little more baryta water, boiling for a few minutes, pouring into a stoppered bottle, and allowing to settle, drawing off the clear liquid, and titrating with standard acid. The exact strengths of the acid solutions must be determined by titration with a standardised alkali solution.

Determine the *heats of neutralisation* by soda of the sulphuric and nitric acids as directed in Part II. Chap. VII. Exp. 6; taking care that the liquids are at the temperature of the air before mixing.

Also determine the thermal value of the reaction which occurs when equivalent masses of nitric acid and sodium sulphate interact.

Convenient quantities of the solutions to use are those which contain 10 grams NaOH, 12·25 grams $H_2SO_4$, 15·75 grams $HNO_3$, and 17·75 grams $Na_2SO_4$, respectively.

Determine also the quantities of heat which disappear when $Na_2SO_4Aq$ and $H_2SO_4Aq$ interact in the ratios of $Na_2SO_4 : \frac{1}{4}H_2SO_4$; $Na_2SO_4 : H_2SO_4$; $Na_2SO_4 : 2H_2SO_4$; and $Na_2SO_4 : 4H_2SO_4$; the results ought to agree fairly with those calculated by the equation

$$[Na^2SO^4Aq,\ nH^2SO^4Aq] = -\frac{n}{n + \cdot8}\,3300.$$

From your results find the relative affinity of sulphuric acid for soda, that of nitric acid being taken as unity.

**Exp. 2.   Chemical method of determining relative affinities of acids.** Read Ostwald, *Chemische Affinitätsbestimmungen*, Journal für prakt. Chem. (2). 22, 251.

Guldberg and Waage's theory of mass-action is assumed. According to this theory, if an acid acts on the salt of another acid, which salt is insoluble in water, with production of a new soluble salt and a new soluble acid, the equation $k.p.q = k'.p'.q'$ applies, except that the term corresponding to the insoluble salt becomes constant. If for example an equivalent of hydrochloric acid ($p$) interacts with an excess of calcium oxalate ($q$), then at all stages of the process calcium chloride ($p'$) and oxalic acid ($q'$) are present in equivalent quantities. By putting the mass of hydrochloric acid originally present as $= 1$, and that of the oxalate dissolved as $= x$, where 1 and $x$ represent equivalents, the equation becomes $k(1-x)c = k'.x^2$; and hence,

$$\frac{k}{k'} = \frac{x^2}{c(1-x)} \text{ and } \sqrt{\frac{k}{k'}} = \frac{x}{\sqrt{c(1-x)}}.$$

The quantity $x$ can be measured directly; $c$ (the chemical mass of the calcium oxalate) is constant.

By repeating the determination of the oxalate dissolved when another acid is used, a second expression is obtained of the form

$$\sqrt{\frac{k'}{k''}} = \frac{x'}{\sqrt{c(1-x')}};$$

where $x' =$ the quantity of oxalate dissolved by the second

12—2

acid.  The  ratio  $\dfrac{x}{\sqrt{1-x}} : \dfrac{x'}{\sqrt{1-x'}}$  gives  the  ratio  of  the
affinities  of  the  two  acids.   (For  more  details  s.  Pattison
Muir's *Principles of Chemistry*, pars. 219, 220.)

For  example,  in  studying  the  reaction  between  calcium
oxalate  and  hydrochloric  and  nitric  acids,  respectively—one
gram-equivalent  of  acid  in  8  litres  of  water—Ostwald  found
that  7·87 p.c.  of  the  oxalate  was  dissolved  when  hydrochloric
acid  was  used,  and  8·01 p.c.  when  nitric  acid  was  used,  at  20°
(i.e.  100  equivalents  of  hydrochloric  acid  dissolved  7·87  equivalents  of  the  oxalate,  and  100  equivalents  of  nitric  acid  dissolved
8·01  equivalents  of  the  oxalate);  hence,  putting  the  acid  as  1,

$x = ·0787$, and $\dfrac{x}{\sqrt{1-x}} = ·0146$; and hence also $x' = ·0801$, and

$\dfrac{x'}{\sqrt{1-x'}} = ·0141$.   The  ratio  ·0146 : ·0141 = 1 : ·97;  hence  the
relative  affinity  of  nitric  acid  is  ·97,  that  of  hydrochloric  acid
being  unity.

Method of procedure.  A quantity of *calcium oxalate*,
$CaC_2O_4.H_2O$, is  prepared  by  mixing  concentrated  warm  solutions  of  *ammonium oxalate* and  *calcium chloride*,  washing  the
pp.  thoroughly,  and  drying  at  100°.   A  solution  of  *hydrochloric
acid* containing  36·37  grams  HCl  in  8  litres  of  water  is  also
prepared;  also  a  *very dilute standardised solution of potassium
permanganate* (·316 grams  in  1000 c.c.  is  a  suitable  strength;
1 c.c. = ·45 mgms. $H_2C_2O_4$).

Arrange  an  apparatus  as  shewn  in  Fig. 55.   *A* is  an  empty
bottle  of  about  500 c.c.  capacity;  *B* is  a  smaller  bottle;  *a* is
an  inverted  thistle-headed  tube  covered  with  muslin,  over
which  is  placed  a  round  piece  of  filter  paper,  which  is  then
covered  with  another  piece  of  muslin;  the  upper  end  of  this
thistle-headed  tube  communicates,  air-tight,  with  a  pipette,  as
shewn.   When  the  air  in  *A* is  compressed  by  means  of  the
india-rubber  ball  the  liquid  in  *B* is  forced  upwards,  through
the  filter  *a*,  into  the  pipette;  any  solid  matter  suspended  in
the  liquid  remains  in  *B*.

Place  a  weighed  quantity  (several  grams)  of  calcium  oxalate
in  *B*,  and  add  a  measured  volume  of  the  solution  of  hydrochloric  acid;  taking  care  that  there  is  considerably  more
calcium  oxalate  than  the  acid  is  capable  of  dissolving.   Keep  *B*
in  water  at  20°  until  the  system  has  attained  equilibrium
(2  hours  or  so  is  sufficient),  shaking  from  time  to  time;  then

compress the air in $A$, and so force liquid through the little filter into the pipette; transfer a pipette-full of the clear

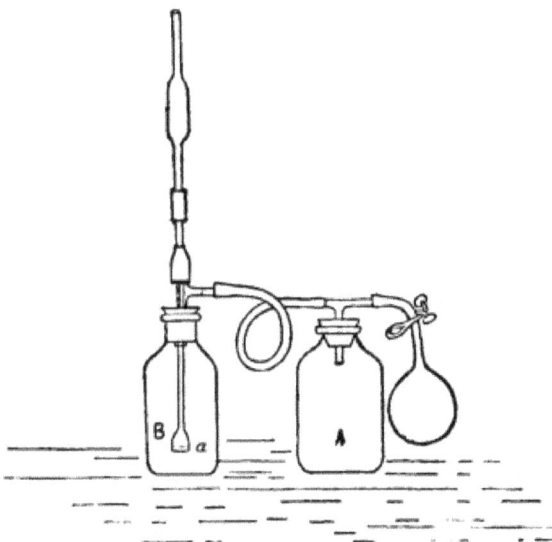

Fig. 55.

liquid to a white basin, and determine the oxalic acid by titration with the dilute standardised permanganate.   Repeat the experiment two or three times.

Arrange and conduct a similar experiment using a solution of *nitric acid* containing 62·89 grams in 8 litres.

From your results deduce the affinity of nitric acid, that of hydrochloric acid being taken as unity.

**Exp. 3.   Volumetric method of determining relative affinities of acids.**   Read Ostwald, *Volumchemische Studien über Affinität*, Journal für prakt. Chem. (2). 16, 385.

When equivalent masses of an acid and a base interact in dilute aqueous solution, a change occurs in the volume of the solution, and it is possible to deduce the amount of chemical change from measurements of this change of volume.   If the same base is then caused to interact with another acid and the volume-change is measured, and if both acids are then allowed to interact with the base, and the volume-change is determined, it is possible to deduce the mass of base which has interacted with each acid to form a salt, and hence to find the relative affinities of the two acids for this base.

Make an aqueous solution of *pure potash* containing 56 grams KOH in 1000 c.c. ; keep this normal solution in a well stoppered bottle. Determine the specific gravity of some *pure concentrated nitric acid*, and deduce the percentage of $HNO_3$ in the liquid; dilute a measured quantity with such a volume of water that 1000 c.c. of the solution contain rather more than 63 grams of $HNO_3$. Now determine the exact strength of this acid by means of the normal potash solution, and then adjust the concentration of the acid solution until 1000 c.c. contain exactly 63 grams of $HNO_3$. By a similar procedure prepare about 1 litre of a *dilute solution of sulphuric acid* of the concentration 98 grams $H_2SO_4$ per 1000 c.c.

Now thoroughly clean a Sprengel's specific gravity tube, by washing it with potash, then with hydrochloric acid, then with water, and finally with alcohol ; dry the apparatus by warming it, and then sucking a current of dry air through it for some time, the air being passed through a plug of cotton wool as shewn in Fig. 56 to prevent the entrance of dust.

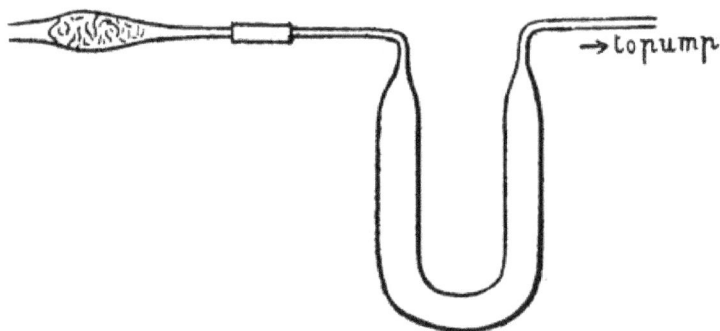

Fig. 56.

Weigh the dry specific gravity tube accurately. Then fill the instrument with water which has been boiled and cooled *in vacuo* to free it from air, and weigh again.

Wash out the apparatus with a little of the normal potash solution, then fill it to the marks with this liquid and weigh again. Proceed similarly with the nitric acid and sulphuric acid solutions.

Now make a mixture of 100 c.c. of the normal potash solution with 100 c.c. of the normal nitric acid solution ; stir

well and allow to stand for some time; then fill the specific
gravity apparatus with the liquid and weigh again. In a
similar way make, and determine the specific gravities of,
mixtures of (1) 200 c.c. normal potash and 100 c.c. sulphuric
acid solutions, (2) 200 c.c. potash with 200 c.c. nitric acid and
100 c.c. sulphuric acid solutions.

Let $x$ equivalents ($x < 1$) of the base (potash) interact
with the nitric acid, when this acid and sulphuric acid are
mixed with the base in equivalent quantities; and let $V$ be
the volume-change when one equivalent of base interacts with
one equivalent of nitric acid. Then $1 - x$ equivalents of the
base must have interacted with the sulphuric acid.

Let $V'$ be the volume-change when one equivalent of the
base interacts with one equivalent of sulphuric acid; and let
$V''$ be the volume-change when equivalent masses of the two
acids, nitric and sulphuric, simultaneously interact with a
quantity of the base equivalent to the mass of one of the acids
present.

Then $V'' = xV + (1 - x) V'$; hence the value of $x$ can be
found.

The ratio $\dfrac{x}{1 - x}$ gives the relative affinities of the two acids.

The following example will serve to shew the accuracy of the method
when worked with a fair amount of care.

Weight of specific gravity tube

$$\text{filled with water at } 20^0 = 26\cdot5712$$
$$\text{alone } \dots\dots\dots\dots\dots = 13\cdot6466,$$
$$\therefore \text{ weight of water filling the tube} = 12\cdot9246.$$

Weight of specific gravity tube filled with

$$
\left.
\begin{array}{ll}
\text{sulphuric acid solution } \dots\dots\dots\dots & = 27\cdot380 \\
\text{nitric acid solution } \dots\dots\dots\dots\dots & = 27\cdot027 \\
\text{potash solution} \dots\dots\dots\dots\dots\dots & = 27\cdot0568 \\
\text{mixed potash and nitric acid} \dots\dots\dots & = 26\cdot9428 \\
\text{potash and sulphuric acid } \dots\dots\dots & = 27\cdot0698 \\
\text{potash, nitric and sulphuric acids} \dots & = 27\cdot0302
\end{array}
\right\} \text{ at } 20^0.
$$

The following specific gravities are deduced from these numbers;—

$$\text{specific gravity of sulphuric acid solution} = 1\cdot0626$$

| | | | | |
|---|---|---|---|---|
| ,, | ,, | ,, nitric acid | $\dots\dots\dots\dots$ | $= 1\cdot0352$ |
| ,, | ,, | ,, potassium sulphate | $\dots$ | $= 1\cdot0386$ |
| ,, | ,, | ,, ,, nitrate | $\dots\dots$ | $= 1\cdot029$ |
| ,, | ,, | ,, potash | $\dots\dots\dots\dots\dots$ | $= 1\cdot0375$ |
| ,, | ,, | ,, mixture of two acids and two salts | | $= 1\cdot0355.$ |

Hence the specific volume—i.e. $\dfrac{1}{\text{sp. grav.}}$ —of the potassium nitrate solution is ·971.

But if no volume-change had occurred, the specific volume of this solution would have been ·965. Hence when equivalent masses of potash and nitric acid interacted in dilute aqueous solution an expansion occurred which can be represented by ·006. Similarly when equivalent masses of sulphuric acid and potash interacted in dilute aqueous solution the expansion which occurred can be represented by ·003; and when equivalent masses of the two acids and potash interacted the expansion which occurred can be represented by ·005.

**Hence**            $·005 = x \times ·003 + ·006\,(1 - x),$
            $\therefore x = \tfrac{1}{3}.$

The observed value of the ratio $x : 1 - x$ shews that the affinity of nitric acid towards potash is double that of sulphuric acid. It is here assumed that the only change which occurs when the two acids and the base are mixed is formation of **the** two normal salts; as a matter of fact the change is complicated by an interaction between the sulphuric acid and the normal potassium sulphate formed. (comp. *Thermal methods of measuring affinities*, in this Chapter.)

# CHAPTER IV.

## METHODS OF DETERMINING THE CONSTITUTIONS OF COMPOUNDS.

Assuming the molecular formulae of a series of compounds to have been determined, the reactions of formation and decomposition of the compounds are studied, and the relations between the compounds established by these reactions are expressed in formulae which rest on the atomic and molecular theory and more especially on the hypothesis of valency.

Determinations of certain physical or physico-chemical constants for series of compounds also sometimes enable conclusions to be drawn concerning the constitutions of these compounds; for instance, determinations of the specific volumes of gasifiable carbon compounds, or of the rates of etherification of alcohols and acids, have thrown light on the constitutions of many compounds.

**Exp. 1. Chemical method of investigating the constitutions of compounds.** (*s.* Pattison Muir's *Principles of Chemistry*, Book I.; especially *Chap. II. sects.* 3 *and* 4.) The molecular formulae of the compounds studied in the following Exps. are assumed to be known. Details of the preparation of the compounds will be found in manuals of organic chemistry; the student should consult one of these manuals, and draw up a plan for the preparation of each compound: only the outlines of the scheme of experimental study are given here.

(1) Prepare absolute *alcohol*, $C_2H_6O$. (2) Dissolve *sodium* in a portion of the *alcohol*; prove that hydrogen is evolved; evaporate the liquid *in vacuo*; prove that the crystals obtained have the composition $C_2H_5ONa.3C_2H_6O$, by heating a weighed quantity to $180°$ and noting the loss of weight (the loss is alcohol), and estimating sodium in the residue. (3) Compare the reactions between (*a*) $PCl_3$ and $H_2O$, and (*b*) $PCl_3$ and $C_2H_6O$.

The results of Exps. (1), (2) and (3) suggest the formula $C_2H_5.OH$ for *alcohol*.

(4) From another portion of the alcohol prepare *aldehyde*, $C_2H_4O$. (5) From another portion of the alcohol prepare *acetic acid* $C_2H_4O_2$. Prove acetic acid to be *monobasic*. (6) From a portion of the acetic acid prepare *acetyl chloride*, $C_2H_3OCl$.

The results of Exps. (4), (5), and (6) suggest the formula $C_2H_3O.OH$ for *acetic acid*.

(7) Mix dry *sodium acetate* with dry *caustic soda*, heat, and prove that *methane*, $CH_4$, is evolved, and *sodium carbonate* remains. (8) From another portion of the acetic acid made in (5) prepare, i. *monochloracetic acid* $C_2H_2ClO.OH$ ; ii. *trichloracetic acid* $C_2Cl_3O.OH$ ; and prove each acid to be *monobasic*.

The results of Exps. (7) and (8) suggest the formula $CH_3.COOH$ for *acetic acid*; and taken along with the former results, they suggest the formulae $CH_3.CH_2OH$ and $CH_3.CHO$ for *alcohol* and *aldehyde*, respectively.

(9) From aldehyde prepare *ethylidene chloride*, $C_2H_4Cl_2$. (10) From ethylene prepare *ethylene chloride*, $C_2H_4Cl_2$. (11) Prove that the ethylidene chloride prepared from aldehyde is different from the ethylene chloride prepared from ethylene, by comparing the boiling points and specific gravities of the two compounds.

Assuming the structural formula of the molecule of ethylene to be $H_2C—CH_2$, and the addition of chlorine to result in the formation of the molecule $ClH_2C—CH_2Cl$, the results of Exps. (9), (10), and (11) confirm the formula $CH_3.CHO$ for *aldehyde*, and hence the formula $CH_3.CH_2OH$ for *alcohol*.

(12) From aldehyde prepare *alcohol* by the action of *nascent hydrogen*.

Exp. (12) confirms the formula $CH_3.CH_2OH$ for alcohol.

Draw up an account of the experiments performed; and state clearly the reasoning employed, the assumptions made, and the conclusions arrived at regarding the constitution of the molecules of alcohol, aldehyde, and acetic acid.

### Exp. 2. Rates of etherification of alcohols by acetic acid.

When an alcohol is heated with acetic acid in molecular proportions a certain amount of the alcohol is changed to an ethereal acetate. The amount of change which occurs in one hour has been called the *initial velocity of etherification*, and the amount which occurs when the whole system has attained equilibrium has been called the *limit of etherification*. The values of these constants for series of alcohol are connected

with the constitutions of the alcohols.    (*s.* Pattison Muir's
*Principles of Chemistry*, Book I. Chap. IV. sect. 4.)

Rates of etherification of ethylic, propylic, and
normal primary butylic, alcohol.    Read, Menschutkin,
*Recherches sur l'influence exercée par l'isomérie des alcools et
des acides sur la formation des Éthers composés.*    Annales de
Chim. et de Phys. (5). 20, 289.

About 5 grams of each *alcohol*; 10 or 15 grams of pure
*acetic acid*; and a *dilute baryta solution* standardised against
acetic acid, are required.

Mixtures of the alcohols with acetic acid in molecular
proportions are heated in very small sealed tubes of about 1
c.c. capacity.    About 4 or 5 grams of each alcohol is placed in
a small stoppered bottle of about 20 c.c. capacity, which has
been previously weighed; the bottle and its contents are
weighed, and the mass of alcohol is thus accurately determined.    The necessary quantity of
acetic acid is calculated, and is then added
from a small burette with a fine opening,
graduated to $\frac{1}{20}$ c.c. ; the weight of a drop of
pure acetic acid delivered by this pipette is
determined, and the exact quantity of acid can
then be delivered by counting the drops after
the greater part of the acid has been run into
the alcohol.

The tubes to contain the mixture are of
the shape shewn in Fig. 57.    A piece of caoutchouc tubing, connected with an india-rubber
ball and furnished with a screw-clamp, is
attached to *a*; the ball is squeezed, and the
other end of the tube is dipped into the liquid,
which is then sucked into the tube by slowly
releasing the ball.    When the little bulb is
about $\frac{2}{3}$ filled, the screw-clamp on the caoutchouc
tubing is closed, and the tube is sealed off at *b*;

Fig. 57.

by gently tapping the little bulb the liquid is caused to rise
above *c*, and the tube is then sealed off at *c*.    The small bottle
containing the mixture of alcohol and acid is again weighed,
and the weight of the mixture in the little bulb is thus
determined.

Four tubes are to be filled with each mixture of alcohol
and acid; making twelve tubes in all.    The tubes are then to
be suspended in a bath of glycerin, placed on a sand-tray, and

heated by means of a Bunsen-lamp furnished with a gas regulator and surrounded by a screen to stop draughts of air. The temperature of the glycerin is raised to $154°$—$156°$ and the tubes are then slowly immersed.

Six tubes, being two of each series, are removed at the end of an hour, and the remaining six are removed after 48 hours. When each tube is cold it is placed in a stoppered bottle containing 30—40 c.c. pure alcohol and 4 or 5 drops of a very dilute alcoholic solution of *rosolic acid*; the tube is broken by agitating the bottle, and the acetic acid which has not been decomposed is determined by the standardised baryta solution.

As the amount of acetic acid originally present is known, it is easy to calculate the percentage of acid, and hence the percentage of alcohol, which has been changed to an ethereal salt. The mean of each pair of experiments after etherification has proceeded for one hour is taken as the initial velocity of etherification, and the mean of each pair after 48 hours as the limit of etherification.

The experiments of Menschutkin proved that the systems consisting of acetic acid and ethylic, propylic, or butylic, alcohol attain their final equilibrium after about 48 hours at $155°$. Menschutkin obtained the following results :—

|  | Initial velocity. | Limit. |
|---|---|---|
| Acetic ethylic system | 46·95 | 66·57 |
| „ propylic „ | 46·92 | 66·85 |
| „ butylic „ | 46·85 | 67·30 |

## Exp. 3.   Specific volumes.

The specific volume of a gasifiable compound is usually defined as $\dfrac{\text{molecular weight}}{\text{spec. grav. of liquid at B. P.}}$.

The specific volume of a solid compound is usually defined as $\dfrac{\text{reacting weight}}{\text{spec. grav. of solid}}$.

Determinations of these constants often throw light on questions regarding the constitutions of compounds. (*s.* Pattison Muir's *Principles of Chemistry*, Book I. Chap. IV. sect. 3.)

Specific volumes of liquids. Read, Ramsay on the *Volumes of liquids at their boiling-points obtained from unit-*

*volumes of their gases.* C. S. Journal, Trans. 1879, p. 463.

Fig. 58 shews the apparatus required. It consists of (1) a small bulb of thin glass of about 10 c.c. capacity, sealed at one end, and terminating at the other end in a capillary tube bent into the form of a hook; (2) a glass vessel of the form, and about three times the size, of that shewn in the Fig. The bulb is suspended by thin platinum wire, as shewn; the exit tube from the glass vessel may be connected with a condenser if necessary.

The bulb is cleaned and dried, and its weight is carefully determined; it is then filled with boiled distilled water at a known temperature, and weighed; from these data, and the known expansion of water, the capacity of the bulb at $0°$ is determined. A little water is then placed in the glass vessel; the bulb (filled with water) is suspended as shewn in the Fig.; the water is boiled, and the bulb is allowed to remain in the steam until drops of water no longer flow from the capillary opening; the source of heat is removed and the bulb is allowed to cool; when cold, it is carefully dried externally, and weighed. The results of this experiment afford data for finding the correction to be applied for the expansion of the glass of the bulb, and for the difference between the temperature of a liquid in the bulb suspended in the vapour obtained by boiling that liquid in the glass vessel and the

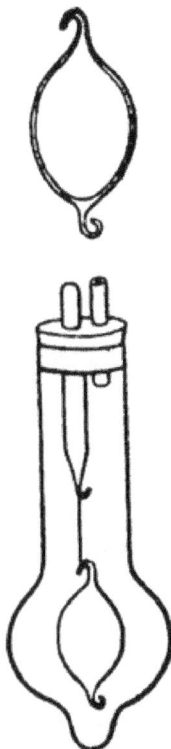

Fig. 58.

true boiling point of the liquid. The calculation is made by finding the volume at $100°$ of the water contained by the bulb at $0°$, from the data of Regnault and Kopp, and comparing this with the observed volume of the water contained by the bulb when heated in steam: the result is stated in the form of a coefficient*.

The bulb is now emptied, by placing its open end downwards and heating, rinsed with alcohol, and dried; a little of the liquid the specific volume of which is to be determined is

* Ramsay found that the coefficient for almost every bulb is ·00015. One volume of water at $0°$ becomes $1·042986$ vols. at $100°$.

now brought into the bulb, by warming and at once plunging
the open end into the liquid; the bulb is rinsed with this
liquid and then emptied, and this process is repeated two or
three times.   The bulb is now nearly filled with the liquid to
be examined, by introducing a little as described, boiling
this by means of a lamp or by surrounding it with hot sand,
and plunging the open end into the liquid, and repeating this
process until sufficient liquid is got into the bulb.   Should the
liquid to be examined be very volatile, or be decomposed by air
or moisture, it is advisable to fill the bulb by suspending it in
the glass vessel, inserting a glass rod in place of the exit tube
from the vessel, and connecting the lower end of the bulb with
this rod by a platinum wire; some of the liquid is then
quickly brought into the glass vessel and boiled; the bulb
being thus heated is lowered into the liquid and tilted, by
means of the two wires and glass rods, so that its neck is
beneath the level of the liquid for a moment; a little liquid
enters; the bulb is raised and the liquid is boiled; this process
is repeated till sufficient liquid has been got into the bulb.

When the bulb has been nearly filled with the liquid whose
specific volume is to be determined, it is suspended in the glass
vessel, in which is placed a little of the same liquid; the liquid
in the vessel is boiled; when drops no longer flow from the
capillary opening of the bulb, the boiling is stopped; the bulb
is allowed to cool, when it is dried and weighed.

The specific gravity of the liquid at its boiling point is
calculated by the formula;—

$$\text{specific gravity} = \frac{W'}{(1 + at)\, W}$$

where $W'$ = weight of liquid in the bulb, $W$ = weight of water
which fills the bulb at $0^\circ$, $t$ = boiling point of liquid, and $a$ = a
coefficient as determined by experiment (usually about ·00015).

The student should determine the specific volumes of (1)
*ethylic alcohol*, (2) a *paraffin* or *olefine* boiling from $40^\circ$ to $90^\circ$,
(3) *benzene*, (4) *phenol*.

Specific volumes of solids.   Read Thorpe and Watts
on the *Specific volumes of water of crystallisation*. C. S. Journal,
Trans. 1880, p. 102.

A stoppered specific gravity bottle of 25 c.c. capacity with
a narrow neck, and several small weighing tubes, must be
provided.   The bottle is carefully cleaned, dried, and weighed,
these processes being repeated two or three times.   It is then

filled to the mark on the neck with *benzene* which has been
purified by freezing, and immersed in water of a known temperature for some hours; the level of the benzene is then accurately
adjusted to the mark on the neck, and the bottle is dried and
weighed. This process is repeated several times. The data
for the bottle are thus obtained.

The compounds whose specific volumes the student is asked
to determine are, *copper sulphate* and *its various hydrates,
cupric hydroxide,* and *cupric oxide.*

(a) *Copper sulphate*; $CuSO_4$. Pure copper sulphate is recrystallised from water. The crystals are powdered and dried,
and a weighed quantity is heated to $280^\circ$, in watch-glasses, until
it ceases to lose weight. The dry salt is then transferred to one
of the weighing tubes and placed in the air bath at $280^\circ$; after
a little time the tube is removed to an exsiccator and allowed
to cool; a few grams of the salt are then *quickly* transferred
to the specific gravity bottle; the bottle and its contents are
heated to $280^\circ$ until the weight is quite constant. The bottle
is filled with benzene, and the necessary weighings are made.
Two independent series of observations should be made.

(b) *Pentahydrated copper sulphate*; $CuSO_4 . 5H_2O$. Prepared by re-crystallising pure copper sulphate from water,
powdering, and drying by pressure between filter paper.

The specific gravity is determined as described, but it is
not necessary to heat the bottle or the salt.

(c) *Trihydrated copper sulphate*; $CuSO_4 . 3H_2O$. Pour a
cold saturated solution of pure copper sulphate into an equal
volume of sulphuric acid of specific gravity 1·7 ; wash the pp.
which forms with small successive quantities of absolute
alcohol until the washings are free from sulphuric acid, and
dry between filter paper. Make determinations of the water,
by drying at $280^\circ$, and the sulphuric acid, by precipitating as
$BaSO_4$.

The specific gravity is determined as in (b).

(d) *Dihydrated copper sulphate*; $CuSO_4 . 2H_2O$. Pour a
cold concentrated solution of pure copper sulphate into
concentrated sulphuric acid with constant stirring; wash with
absolute alcohol, and dry between filter paper. Make determinations of the water and sulphuric acid.

The specific gravity is determined as in (b).

(e) *Monohydrated copper sulphate*; $CuSO_4 . H_2O$. Heat the
pentahydrate to $110^\circ$ until it ceases to lose weight at that temperature. Make a determination of sulphuric acid, or of water.

Determine the specific gravity as in (*a*), heating the specific gravity bottle with the salt in it to 110° until the weight is constant.

(*f*) *Cupric hydroxide*; $CuO_2H_2$. Add potash to a cold rather dilute solution of pure copper sulphate until the blue colour of the liquid has nearly disappeared; collect the pp. on a filter, wash it as rapidly as possible with cold water until the washings are free from sulphates, dry by pressure between filter paper and then over sulphuric acid. Make determinations of water by heating to redness, and of copper by *standardised potassium cyanide* solution.

The specific gravity is determined as in (*b*).

(*g*) *Cupric oxide*; CuO. Heat a portion of the hydroxide (*f*) to 150°—200° until it ceases to lose weight. Determine the specific gravity as in (*b*).

Assuming that the differences between the specific volumes of the various hydrated copper sulphates represent the specific volumes of the *water of crystallisation* combined with copper sulphate to form these various hydrates, and that the difference between the specific volumes of CuO and $CuO_2H_2$ represents the specific volume of the *water of constitution* which chemically interacts with CuO to form $CuO_2H_2$, compare the results you have obtained in exps. (*a*) to (*e*) with those obtained in exps. (*f*) and (*g*), and shew how they help to establish a distinction between *water of crystallisation* and *water of constitution*. As additional data in this comparison make use of the following specific gravities of oxides and hydroxides or hydrated oxides ;—

| | $I_2O_5$ | $I_2O_5.H_2O$ | | CaO | $CaO.H_2O$ | | $B_2O_3$ | $B_2O_3.3H_2O$ |
|---|---|---|---|---|---|---|---|---|
| *Spec. Grav.* | 4·49 | 4·27 | | 3·18 | 2·1 | | 1·81 | 1·44 |

Also use the following specific gravities of crystalline salts and hydrated salts ;

| | $CaCl_2$ | $CaCl_2.6H_2O$ | | $BaCl_2$ | $BaCl_2.2H_2O$ | | $Na_2B_4O_7$ | $Na_2B_4O_7.10H_2O$ |
|---|---|---|---|---|---|---|---|---|
| *Spec. Grav.* | 2·24 | 1·635 | | 3·886 | 3·052 | | 2·367 | 1·692 |

Calculate in each case the mean specific volume of each molecule of *water of crystallisation*, and compare this volume with the mean specific volume of each molecule of *water of constitution*.

The following are the values obtained by Thorpe and Watts for the spec. volumes of copper sulphate and its different hydrates ;—M = $CuSO_4$.

| | M | $M.H_2O$ | $M.2H_2O$ | $M.3H_2O$ | $M.5H_2O$ |
|---|---|---|---|---|---|
| Spec. volumes | 44·1 | 54·6 | 66·1 | 80 | 109 |
| *Differences.* | | 10·5 | 11·5 | 13·9 | 29 |

# APPENDIX I.

I. GIVEN two elements. Find what elements they are; then prepare a compound of each of these elements and prove experimentally that the bodies you have prepared are compounds.

*(Ferrum redactum; Sulphur.)*

II. Of the two bodies A and B, one is an element and the other is a compound. Examine quantitatively the action of heat on A and B, and also the action of dilute sulphuric acid on A. From the results of your experiments determine as far as you can which is the element and which is the compound.

*(A = Ferrum redactum; B = Potassium chlorate.)*

III. Given three oxides A, B, and C. Determine experimentally which is a basic oxide, which is an acidic oxide, and which is a peroxide.

*(A = Manganese dioxide; B = Chromic oxide, $Cr_2O_3$; C = Chromic anhydride.)*

IV. Given two elements. From the properties of their oxides classify the elements as metals or non-metals.

*(Zinc powder; Flowers of sulphur.)*

V. Given four elements. Prepare an oxide of each, and examine the interaction of water with each oxide, determining whether an acid or an alkali is formed, or whether no change occurs. Then act on each element with dilute sulphuric acid and determine whether gas is evolved; in cases of gas-evolution find what particular gas is given off. In those cases where an action occurs and the element dissolves, boil down the liquids, and find whether salts have been formed. From the results obtained classify the four elements as metals and non-metals. What other properties would you expect to belong to those which are metals, and what other properties to those which are non-metals?

(*Ferrum redactum; Flowers of sulphur; Powdered magnesium; and Powdered charcoal.*)

VI. Prove that the given oxide is not acidic, but that it reacts with an acid to form a salt.

(*Chromic oxide*, $Cr_2O_3$.)

VII. Given four salts, A, B, C, and D. Sulphuric acid reacts normally with two of these salts, A and B: the products of the interaction of this acid with C and D are not such as would result from a normal reaction between an acid and a salt. Prove these statements experimentally.

(A = *Potassium chloride*; B = *Lead acetate*; C = *Potassium iodide*; D = *Mercuric chloride*.)

VIII. Given red lead. From it prepare two other oxides of lead. Prove one to be a basic oxide and the other a peroxide. Determine experimentally whether the peroxide does or does not react under any conditions as an acidic oxide.

IX. Given a salt A. Find its qualitative composition; then prove that the gas evolved when A interacts with concentrated sulphuric acid is soluble in water and that this solution reacts as an acid. By neutralising this acid solution, obtain the original salt in a solid form.

Perform similar experiments with the salt B, and compare the results obtained in this case with those obtained in the case of A.

(A = *Potassium chloride*; B = *Potassium nitrate*.)

X. Given an aqueous solution of an acid, a metallic oxide, a metallic carbonate, and a metal.

Determine qualitatively the change that occurs when (1) the oxide, (2) the carbonate, and (3) the metal, reacts with the acid.

(*Zinc; Magnetic oxide of iron; Sodium carbonate; Hydrochloric acid.*)

XI.  Compare the reactions of the given metal with each
of the acids given; determining, as far as can be done by
qualitative examination, (1) what gas (or gases) is evolved in
each case, (2) what are the compositions of the non-gaseous
products of the reactions.

*(Zinc ; Hydrochloric acid ; Nitric acid.)*

XII.  From the metal A prepare two salts each composed
of the metal and the same acid radicle, establishing by
experiments that the salts you have prepared are different in
their properties.  Find out as much as you can by qualitative
experiments regarding the differences between the compositions
of the two salts.

Determine experimentally whether the metal B forms two
salts with the same acid radicle or only one.

*(A = Iron ; B = Zinc.)*

XIII.  From the given salt, prepare that oxide of the
metal of the salt which contains the metal and oxygen united
in the proportion of three atoms of metal to four atoms of
oxygen.

[You are not required to prove the composition of the oxide
analytically, but merely to identify your preparation with the
oxide $M_3O_4$ from knowing the properties of that oxide.]

Determine by experiment what salts are formed by the
interaction of hydrochloric acid with the oxide you have prepared.  Contrast this reaction with that between the same
acid and the oxide $M_2O_3$ of the same metal, which oxide you
must prepare.

*(Ferrous sulphate crystals.)*

XIV.  Given a salt A and an oxide B.  What changes
occur when each is heated?  Compare the results obtained
with the action of heat on the oxide C.

From an aqueous solution of A prepare the same body as
results from the action of heat on A.

*(A = Potassium chlorate ; B = Mercuric oxide ; C = Litharge.)*

XV.  Given potassium dichromate.  Prepare from it,
   (1)  a solution containing a salt of chromium ;
   (2)  a specimen of pure chromic hydrate ;
   (3)  a specimen of chromic anhydride.

Prepare potassium dichromate from a portion of the chromic
hydrate made in (2).

13—2

XVI. Given manganese sulphate. Prepare therefrom a solution of potassium manganate; transform this solution into one of potassium permanganate, and from this solution pass back again to manganese sulphate.

Prove that an acidulated solution of potassium permanganate oxidises ferrous to ferric salts, and oxalic acid to carbon dioxide.

XVII. Given two oxides, A and B, of the same metal. Prove distinctly and conclusively, (1) that A is an acidic oxide but that B is not acidic ; (2) that A reacts with hydrochloric acid to form a salt ; (3) that B also reacts with the same acid to form a salt.

Determine by experiment what body is produced during the reaction of A with hydrochloric acid that is not produced when B reacts with the same acid.

From A prepare B ; and from B prepare the potassium salt of the acid corresponding to A.

      ($A = Chromic\ anhydride$ ;   $B = Chromic\ oxide$, $Cr_2O_3$.)

XVIII. Given an acid in aqueous solution. Prove experimentally that it is monobasic. [At a red heat this acid is completely volatilised but its potassium salt is unchanged.]

                 (*Hydrochloric acid.*)

XIX. Given solutions of sulphuric acid and barium chloride of stated strengths. Prove that the equation

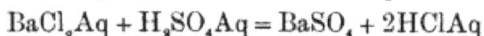

$$BaCl_2Aq + H_2SO_4Aq = BaSO_4 + 2HClAq$$

accurately represents the mutual reaction of these two bodies.

XX. Given phosphorus pentoxide. From it make

      (1)   a solution of metaphosphoric acid ;
and     (2)   a solution of orthophosphoric acid ;
recognising each acid by the usual tests.

From the orthophosphoric acid obtained prepare the salt disodic orthophosphate, and from this prepare a solution of pyrophosphoric acid.

XXI. Prepare oxygen, nitrous oxide, and nitric oxide ; and distinguish these three gases by as many tests as possible.

XXII. From copper sulphate prepare the two oxides of copper. Prepare also the two chlorides of copper, corresponding in composition to the two oxides.

# APPENDIX II.

(more difficult than those in Appendix I.).

I. A and B are aqueous solutions of salts which readily give up oxygen. Find the oxidising power of A; that of B being unity.

(A = *Solution of potassium dichromate;* B = *Solution of potassium permanganate.*)

II. You are given solutions of barium chloride and ammonium oxalate of stated strengths. Determine the influence which variation of temperature exerts on the amount of chemical change which takes place in five minutes, when the solutions are so mixed that molecular proportions of the two salts interact.

III. Determine the equivalents of the three metals A, B, and C, by finding the mass of hydrogen produced by the reaction between a weighed mass of each of the elements and excess of hydrochloric acid.

(*Magnesium; Aluminium; Zinc.*)

IV. Determine the number of c.c. of each of the three aqueous solutions of acids given which contain masses of the acids that are equivalent as regards their reactions with an alkali.

(*Sulphuric acid; Nitric acid; Phosphoric acid.*)

V. Prove by at least three roughly quantitative experiments that iron is an element and not a compound.

Prove that copper oxide is a compound.

State the assumptions made in the reasoning based on the results of your experiments.

VI.   From metallic iron prepare crystals of pure ferrous sulphate.   Convert these crystals into the higher sulphate, and determine by quantitative experiments the chemical change which occurs.

VII.   Given two metallic oxides, A and B.   (1) Determine of what metals they are oxides.   (2) Classify each as salt-forming, acid-forming, or peroxide.   (3) From each prepare an aqueous solution of a salt of potassium in which the metal of the given oxide forms a portion of the non-metallic or acidic radicle ; then act on the solution thus prepared from the oxide A with carbon dioxide, and on the solution prepared from the oxide B with dilute sulphuric acid, and evaporate these solutions to their crystallising points.   What salts are thus obtained?   Examine the reaction of each of these salts with easily oxidised bodies such as oxalic acid and ferrous sulphate in presence of sulphuric acid.

(A = *Manganese dioxide* ;   B = *Chromic oxide* $Cr_2O_3$.)

VIII.   Given sulphur ; prepare a solution containing ·005 gram sulphuric acid per c.c.

IX.   Given mercury, zinc, and copper.   Prepare as many oxides and chlorides as you can from each of these three elements.   From the number of oxides in each case, their compositions, and the action of heat on the several oxides, as well as from the properties of the chlorides corresponding to these oxides, determine whether the three metals, mercury, zinc, and copper, ought to be placed in the same class.

X.   Expired air is said to contain about four per cent. more carbon dioxide, and four per cent. less oxygen, than inspired air.   Put this statement to experimental test.

XI.   Find quantitatively what relations exist between the atomic weights of calcium, strontium, and barium, and the stabilities of the carbonates of these metals with respect to heat.

XII.   "Chemical equations, though right enough as such, as theories of the processes of titration are only approximately correct." Experimentally support this statement by determining iron in a solution of ferric chloride, (1) with thiosulphate of sodium standardised against pure iodine, (2) with thiosulphate of sodium standardised against iodine liberated in the cold from an acid solution of a known mass of potassium dichromate to which excess of potassium iodide has been added.

XIII. Prove that air is a mixture and not a compound of oxygen and nitrogen. This will involve, (1) the proof that oxygen and nitrogen are not present in air in atomic proportions; (2) the proof that when oxygen and nitrogen are mixed in the proportion in which they exist in air there is no evidence of chemical change, but that nevertheless the mixture has the properties of air; (3) the proof that "air" dissolved by water has a different composition from that of the air merely in contact with the water; (4) the proof that nitrogen and oxygen in air can be partially separated by atmolysis.

XIV. Determine the coefficient of solubility of sulphuretted hydrogen in water.

# APPENDIX III.

A. *Detection of commonly occurring metals when not more than one is present in any group.*

I. Test for an ammonium salt by boiling a portion of *original* with excess of *potash* and examine gas for *ammonia*.

II. GROUP I. To original solution add a little *hydrochloric acid* (if no pp. forms Group I. is absent):—

| Pp. forms; add more HClAq | | | |
|---|---|---|---|
| Pp. dissolves, probably a salt of As or Sb. | Pp. does not dissolve; **add** *much cold water* | | |
| | Pp. dissolves, probably Ba or Sr, as chlorides. | Pp. does not dissolve; *boil with much water* | |
| | | Pp. dissolves, Pb($PbCl_2$).* | Pp. does not dissolve; add *ammonia* |
| | | Pp. dissolves, Ag(AgCl). | Pp. is blackened, $Hg^{ous}$(HgCl). / Pp. is unchanged, $SiO_2$. |

III. GROUP II. Into filtrate from Gr. I., or into original liquid (acidulated by HCl) if Gr. I. is absent, pass *sulphuretted*

---

* The formulae in brackets shew the compositions of the pps.; thus if lead were present the pp. produced by HClAq would be $PbCl_2$, this salt is soluble in boiling water.

*hydrogen* till liquid smells strongly of this gas; then, if pp. forms, warm, dilute, warm again and filter; boil filtrate and saturate it with $H_2S$ gas; collect any pp. which forms on same filter as before; again boil filtrate and pass in $H_2S$; and *repeat this treatment until $H_2S$ ceases to produce any change in the filtrate.* Wash pp. thoroughly :—

(*a*)  Pp. produced by $H_2S$ is black or dark brown ($HgS$, $CuS$, $PbS$, $SnS$, or $Bi_2S_3$);

Digest pp. with warm *yellow ammonium sulphide*

| Pp. dissolves, $Sn^{ous}$. | Pp. does not dissolve; wash pp. and warm it with dilute *nitric acid* | | | |
|---|---|---|---|---|
| | Pp. does not dissolve, $Hg^{ic}$. | Pp. dissolves; boil liquid till $H_2S$ is removed, and then add *alcohol and dilute sulphuric acid* | | |
| | | White pp., $Pb(PbSO_4)$. | No pp.; add *excess of ammonia* | |
| | | | White pp., $Bi(BiO_3H_3)$. | Blue liquid, $Cu$. |

(*b*)   Pp. produced by $H_2S$ is yellow ($CdS$, $As_2S_3$, or $SnS_2$);

Digest pp. with warm *yellow ammonium sulphide*

| Pp. dissolves, shews As or $Sn^{ic}$; warm another part of pp. by $H_2S$ with concentrated *solution of ammonium carbonate* | | Pp. does not dissolve, $Cd$. |
|---|---|---|
| Pp. dissolves, As. | Pp. does not dissolve, $Sn^{ic}$. | |

(*c*)   Pp. produced by $H_2S$ is orange-red ($Sb_2S_3$), $Sb$.

Confirm by proving pp. to be soluble in warm $(NH_4)_2S_2Aq$.

IV.  GROUP III.  Boil filtrate from Gr. II., after adding a little $HNO_3Aq$, till liquid no longer smells of $H_2S$ ; add a little more $HNO_3Aq$ and boil again*, then add a considerable quantity of *ammonium chloride* and then *ammonia in very slight excess*† :—

| White pp.; add more NH₃Aq | | | | Brown pp. | Pale green- | Black pp. |
|---|---|---|---|---|---|---|
| | | | | $Fe(Fe_2O_6H_6)$. | ish white pp. | Shews that $H_2S$ was not |
| Pp. dis- solves, probably Zn (*s. next table*) | Pp. does not dissolve; add *potash* | | | | $Cr(Cr_2O_6H_6)$ gives green colour in borax bead. | completely removed be- fore adding NH₃Aq. |
| | Pp. dissolves, $Al(Al_2O_6H_6)$. | Pp. does not dissolve, phos- phate or oxalate of alkaline earth [or per- haps silica] (*s. App. to this table*). | | | | |

*Appendix to Group III.: to test for an alkaline phosphate or oxalate :—*

(*a*)  Test original for, (1) phosphoric acid by *ammonium molybdate* test ; (2) oxalic acid by *calcium chloride* test, after boiling with large excess of *solid sodium carbonate* for some time, to remove metals as carbonates, filtering, and acidulating filtrate with acetic acid (*s. Table for detection of acids*).

(*b*)  Test a portion of filtrate from Gr. II. for barium and strontium by *calcium sulphate* test ; if these metals are absent, test another portion of same liquid for calcium by *sulphuric acid and alcohol* test (*s. Table for Gr. V.*). If none of these metals is found, fuse a portion of *original solid* with $Na_2CO_3$, boil fused mass in water, filter, *wash residue on filter 3 or 4 times with warm water*, then dissolve residue in HClAq, add excess of $NH_4ClAq$ and $NH_3Aq$, (filtering if pp. forms)

---

* The reasons for this treatment are, (1) to remove $H_2S$ ; (2) to oxidise iron salts from ferrous to ferric, the treatment with $H_2S$ in pptg. Gr. II. reduces ferric to ferrous salts.

† Should HCl and $H_2S$ have produced neither a pp. nor a change of colour, it is better to add $NH_4ClAq$ and $NH_3Aq$ to a portion of the origi- nal liquid to which the reagents for Grs. I. and II. have not been added.

and test for magnesium by *sodium phosphate* test* (*s. Table for Gr. VI.*).

V. GROUP IV. To filtrate from Gr. III. add a few drops of *ammonium sulphide* :—

| White pp. appearing greenish in the yellow liquid, Zn(ZnS). | Buff coloured pp., Mn(MnS) gives amethyst colour in borax bead, O. F. | Black pp. (NiS or CoS); wash and examine in borax bead in O. F. | |
|---|---|---|---|
| | | Blue bead, Co. | Reddish bead, Ni. |

VI. GROUP V. To filtrate from Gr. IV. add *ammonium carbonate*, collect pp., wash it, dissolve in HClAq, and divide liquid into two parts :—

| To one part add *calcium sulphate* solution | | If Ba and Sr are absent, to other part add *dilute sulphuric acid and alcohol* |
|---|---|---|
| White pp. forms at once, Ba(BaSO$_4$). | White pp. forms on standing and warming, Sr(SrSO$_4$). | White pp. forms rather slowly, Ca(CaSO$_4$). |

VII. GROUP VI. Divide filtrate from Gr. V. into two parts :—

| To one part add *sodium phosphate:* white pp. forms slowly, Mg(Mg.NH$_4$.PO$_4$). | Evaporate other portion just to dryness, and bring a little of the residue, on Pt wire, into a *Bunsen-flame* | |
|---|---|---|
| | Flame is coloured pale violet visible through blue glass, K. Confirm by dissolving some of residue in a *very little* water, adding conc. solution of sodium-hydrogen tartrate and stirring : white pp. of KHC$_4$H$_4$O$_6$ forms. | Flame is coloured deep yellow, nearly invisible through blue glass, Na. |

* The Mg phosphate or oxalate is decomposed by the Na$_2$CO$_3$, MgCO$_3$ is formed and Na phosphate or oxalate dissolves in the wash-water.

B.  *Detection of commonly occurring acids when not more than one is present in any group\*.*

I.  Test for n i t r i c  a c i d  by *ferrous sulphate* test (*s. Part I. Chap. VII. Exp.* 4).

II.  GROUP I.  To original *solid* (if liquid is given evaporate to dryness on steam-bath) add concentrated *sulphuric acid :—*

| Colourless non-inflammable gas ($CO_2$) evolved; gas produces white pp. in *lime water,* $H_2CO_3$. | Colourless strongly fuming gas evolved, probably HCl (confirm; *s. Gr. IV.*) | Colourless badly-smelling gas evolved, and generally white solid (S) deposited, $H_2S$ (confirm; *s. Gr. III.*) |
|---|---|---|

If no reaction occurs with sulphuric acid in the cold, then heat :—

| Inflammable gas (CO) evolved | | Brown to red gas evolved ($NO_2$ or Br) $HNO_3$ or HBr (confirm; *s. Gr. IV also s. I. above.*) | Reddish gas (III and I) evolved, and violet solid (I) formed. III (confirm; *s. Gr. III.*) | Gas ($SO_2$) smelling of burning S and turning $K_2Cr_2O_7$ Aq greenish | |
|---|---|---|---|---|---|
| No colour change, $H_2C_2O_4$ (confirm; *s. Gr. II.*) | Colour changes from yellow or red to greenish blue, $H_4FeCy_6$ or $H_3FeCy_6$ (confirm; *s. Gr. III.*) | | | With deposition of yellow S, $H_2S_2O_3$. | Without deposition of S, $H_2SO_3$. |

*To prepare the solution for the remaining groups,* add a l a r g e  e x c e s s  of solid *sodium carbonate* to the original liquid, or if original is a solid add $Na_2CO_3$ and a little water†; boil for 5—10 mins., filter off the pp. of metallic carbonates, *just acidify* the filtrate by adding d i l u t e  *nitric acid,* boil, and filter again if necessary. *The solution should now be neutral;* if not, make it so by very cautious addition of *very dilute* ammonia.

III.  GROUP II.  To a *portion* of the n e u t r a l liquid add *barium nitrate* solution ; if a white pp. forms add to it dilute *nitric acid :—*

\* Chromic, arsenic, and arsenious, acids, if present, will have been detected in testing for metals.
† If the addition of excess of $Na_2CO_3$Aq to a little of the original liquid produces no pp. the treatment with $Na_2CO_3$ described above is unnecessary; *the liquid must however be neutral before examining for Grs. II. and III.*

Treatment with $Na_2CO_3$ removes metals, that would interfere with the subsequent tests for acids, by pptg. them as carbonates; the filtrate contains sodium salts of the acids.

| Pp. is undissolved, $H_2SO_4$; or only partly dissolved with evolution of $SO_2$, $H_2SO_3$ containing $H_2SO_4$ (*comp. Gr. I.*) | Pp. dissolves with effervescence, $H_2CO_3$ (*s. Gr. I.*) | Pp. dissolves without effervescence | | |
|---|---|---|---|---|
| | | To original liquid add *calcium chloride* and *acetic acid* | | |
| | | White pp. $(CaC_2O_4)$ shews $H_2C_2O_4$. | No pp.; to another part of original liquid add excess of *hydrochloric acid*, soak piece of *turmeric paper* in liquid for some time and dry it; turmeric turns reddish, shews $H_2B_2O_4$. | No pp. and boric acid absent; to another part of original liquid add *nitric acid*, and *ammonium molybdate*, and warm; yellow pp. shews $H_3PO_4$. |

IV.  GROUP III.  To a *portion* of the neutral liquid add neutral *ferric chloride* solution :—

| Blue pp. $(Fe_7Cy_{18})$  $H_4FeCy_6$. | Black pp. $(FeS)$  $H_2S$. | Brown pp. $(Fe_2O_3)$ probably HCN (*s. Gr. IV.*) | No pp. but production of colour ranging from red-brown to deep claret; add *hydrochloric acid* | | |
|---|---|---|---|---|---|
| | | | Colour is removed and liquid remains colour of dilute $Fe_2Cl_6Aq$, $H_4C_2O_2$. | Colour remains unchanged; to orig. liquid add freshly prepared *ferrous sulphate* solution | |
| | | | | Blue pp. $(Fe_6Cy_{12})$  $H_3FeCy_6$ | No pp.; to original liquid add *carbon disulphide* and a little *chlorine water* and shake well |
| | | | | | CS$_2$ is coloured violet (I), / III.    CS$_2$ remains uncoloured, / HCNS. |

V.  GROUP IV.  To two or three drops (not more) of *silver nitrate* solution in a very clean test tube add a drop or two of the original liquid :—

| White pp. forms, but dissolves on adding more of original liquid, HCN (pp.=AgCN). Confirm by prussian-blue test (*s. Manual of analysis.*) | White to yellow-white pp. forms and does not dissolve on adding more of original liquid:— To original liquid add *carbon disulphide* and a little *chlorine water* and shake well | |
|---|---|---|
| | CS$_2$ is coloured brown (Br), HBr. | CS$_2$ remains uncoloured, HCl. |

# APPENDIX IV.

## 1.  *Atomic weights of the Elements.*

| Element | At. Wt. | Element | At. Wt. | Element | At. Wt. |
|---|---|---|---|---|---|
| Aluminium | 27·02 | Hydrogen | 1 | Ruthenium | 104·4 |
| Antimony | 120 | Indium | 113·4 | Scandium | 44 |
| Arsenic | 74·9 | Iodine | 126·53 | Selenion | 78·8 |
| Barium | 136·8 | Iridium | 192·5 | Silicon | 28·3 |
| Beryllium | 9·08 | Iron | 55·9 | Silver | 107·66 |
| Bismuth | 208 | Lanthanum | 138·5 | Sodium | 23 |
| Boron | 10·9 | Lead | 206·4 | Strontium | 87·3 |
| Bromine | 79·75 | Lithium | 7·01 | Sulphur | 31·98 |
| Cadmium | 112 | Magnesium | 24 | Tantalum | 182 |
| Caesium | 132·7 | Manganese | 55 | Tellurium | 125 |
| Calcium | 39·9 | Mercury | 199·8 | Thallium | 203·64 |
| Carbon | 11·97 | Molybdenum | 95·8 | Thorium | 231·8 |
| Cerium | 139·9 | Nickel | 58·6 | Tin | 117·8 |
| Chlorine | 35·37 | Niobium | 94 | Titanium | 48 |
| Chromium | 52·4 | Nitrogen | 14·01 | Tungsten | 183·6 |
| Cobalt | 59 | Osmium | 193 (?) | Uranium | 240 |
| Copper | 63·2 | Oxygen | 15·96 | Vanadium | 51·2 |
| Didymium | 144 | Palladium | 106·2 | Ytterbium | 173 |
| Erbium | 166 | Phosphorus | 30·96 | Yttrium | 89·6 |
| Fluorine | 19·1 | Platinum | 194·3 | Zinc | 64·9 |
| Gallium | 69 | Potassium | 39·04 | Zirconium | 90 |
| Germanium | 72·3 | Rhodium | 104 | | |
| Gold | 197 | Rubidium | 85·2 | | |

## II.  *Equivalence between English and Metric weights and measures.*

| | | | Logarithms. | Ar. Co. Log. |
|---|---|---|---|---|
| 1 Kilometre | = | 0·6214 Mile. | 9·7933 712 | 0·2066 188 |
| 1 Metre | = | 3·2809 Feet. | 0·5159 930 | 9·4840 070 |
| 1 Centimetre | = | 0·3937 Inch. | 9·5951 742 | 0·4048 258 |

APP. IV.]        NUMERICAL TABLES.        **207**

| | | | Logarithms. | Ar. Co. Log. |
|---|---|---|---|---|
| 1 Cubic Metre | = 35·31660 Cubic Feet. | | 1·5479 790 | 8·4520 210 |
| 1 Cubic Decimetre | = 61·02709 Cubic Inches. | | 1·7855 226 | 8·2144 774 |
| 1 Cubic Centimetre | = 0·06103 ,, ,, | | 8·7855 226 | 1·2144 774 |
| 1 Litre | = 0·22017 Gallon. | | 9·3427 581 | 0·6572 419 |
| 1 Litre | = 0·88066 Quart. | | 9·9448 083 | 0·0551 917 |
| 1 Litre | = 1·76133 Pints. | | 0·2458 407 | 9·7541 593 |

| | | | Logarithms. | Ar. Co. Log. |
|---|---|---|---|---|
| 1 Kilogram | = 2·20462 Pounds Avoirdupois. | | 0·3433 337 | 9·6566 663 |
| 1 ,, | = 2·67922 ,, Troy. | | 0·4280 083 | 9·5719 917 |
| 1 Gram | = 15·43235 Grains. | | 1·1884 321 | 8·8115 679 |

III. *Spec. gravities and compositions of aqueous solutions of Acids, Alkalis, and Alcohol.*

| | Grams $H_2SO_4$ in | | | Grams $HNO_3$ in | | | Grams HCl in | |
|---|---|---|---|---|---|---|---|---|
| S.G. | 100 gm. | 100 c.c. | S.G. | 100 gm. | 100 c.c. | S.G. | 100 gm. | 100 c.c. |
| 1·842 | 100 | 184·2 | 1·530 | 99·84 | 152·75 | 1·212 | 42·9 | 52·0 |
| 1·796 | 86·5 | 155·4 | 1·529 | 99·52 | 152·2 | 1·210 | 42·4 | 51·3 |
| 1·753 | 81·7 | 143·2 | 1·514 | 95·27 | 144·2 | 1·205 | 41·2 | 49·6 |
| 1·711 | 78·1 | 133·6 | 1·506 | 93·01 | 139·1 | 1·199 | 39·8 | 47·7 |
| 1·672 | 74·7 | 124·8 | 1·494 | 89·56 | 133·8 | 1·195 | 39·0 | 46·6 |
| 1·634 | 71·6 | 117·0 | 1·486 | 87·45 | 129·9 | 1·190 | 37·9 | 45·0 |
| 1·597 | 68·6 | 109·5 | 1·482 | 86·17 | 127·7 | 1·185 | 36·8 | 43·6 |
| 1·563 | 65·5 | 102·4 | 1·463 | 80·96 | 118·4 | 1·180 | 35·7 | 42·1 |
| 1·530 | 62·5 | 95·6 | 1·438 | 74·01 | 106·4 | 1·175 | 34·7 | 40·8 |
| 1·498 | 59·6 | 89·3 | 1·432 | 72·39 | 103·7 | 1·171 | 33·9 | 39·7 |
| 1·468 | 56·9 | 83·5 | 1·429 | 71·24 | 101·8 | 1·166 | 33·0 | 38·5 |
| 1·438 | 54·0 | 77·7 | 1·419 | 69·20 | 98·2 | 1·161 | 32·0 | 37·2 |
| 1·410 | 51·2 | 72·2 | 1·400 | 65·07 | 91·1 | 1·157 | 31·2 | 36·1 |
| 1·383 | 48·3 | 66·8 | 1·381 | 61·21 | 84·5 | 1·152 | 30·2 | 34·8 |
| 1·357 | 45·5 | 61·7 | 1·372 | 59·59 | 81·8 | 1·143 | 28·4 | 32·5 |
| 1·332 | 43·0 | 57·3 | 1·353 | 56·10 | 75·9 | 1·134 | 26·6 | 28·8 |
| 1·308 | 40·2 | 52·6 | 1·331 | 52·33 | 69·6 | 1·125 | 24·8 | 27·9 |
| 1·285 | 37·4 | 48·1 | 1·323 | 50·99 | 67·5 | 1·116 | 23·1 | 25·8 |
| 1·263 | 34·7 | 43·8 | 1·298 | 47·18 | 61·2 | 1·108 | 21·5 | 23·8 |
| 1·241 | 32·2 | 40·0 | 1·274 | 43·53 | 55·5 | 1·100 | 19·9 | 21·9 |
| 1·220 | 29·6 | 36·1 | 1·237 | 37·95 | 46·9 | 1·091 | 18·1 | 19·7 |
| 1·200 | 27·1 | 32·5 | 1·211 | 33·86 | 41·0 | 1·083 | 16·5 | 17·9 |
| 1·180 | 24·5 | 28·9 | 1·172 | 28·00 | 32·8 | 1·075 | 15·0 | 16·1 |
| 1·162 | 22·2 | 25·8 | 1·157 | 25·71 | 29·8 | 1·067 | 13·4 | 14·3 |
| 1·142 | 19·6 | 22·4 | 1·105 | 17·47 | 19·3 | 1·060 | 12·0 | 12·7 |
| 1·125 | 17·3 | 19·5 | 1·067 | 11·41 | 12·2 | 1·052 | 10·4 | 10·6 |
| 1·108 | 15·2 | 16·8 | 1·045 | 7·72 | 8·1 | 1·044 | 8·9 | 9·3 |
| 1·091 | 13·0 | 14·2 | | | | 1·036 | 7·3 | 7·6 |
| 1·075 | 10·8 | 11·6 | | | | 1·029 | 5·8 | 6·0 |
| 1·060 | 8·8 | 9·3 | | | | 1·022 | 4·5 | 4·6 |
| 1·045 | 6·8 | 7·1 | | | | 1·014 | 2·9 | 2·9 |

| KOH in | | | NaOH in | | | Alcohol in | | |
|---|---|---|---|---|---|---|---|---|
| S.G. | 100 gm. | 100 c.c. | S.G. | 100 gm. | 100 c.c. | S.G. at 15° C. | 100 gm. | 100 c.c. |
| 1·790 | 70 | 125·30 | 1·748 | 70 | 122·36 | ·7947 | 100 | 79·47 |
| 1·729 | 65 | 112·38 | 1·695 | 65 | 110·18 | ·8093 | 95 | 76·88 |
| 1·667 | 60 | 100·02 | 1·643 | 60 | 98·58 | ·8232 | 90 | 74·09 |
| 1·604 | 55 | 88·22 | 1·591 | 55 | 87·51 | ·8363 | 85 | 71·08 |
| 1·539 | 50 | 76·95 | 1·540 | 50 | 77·00 | ·8488 | 80 | 67·90 |
| 1·475 | 45 | 66·38 | 1·488 | 45 | 66·96 | ·8610 | 75 | 64·58 |
| 1·412 | 40 | 56·44 | 1·437 | 40 | 57·48 | ·8729 | 70 | 61·10 |
| 1·349 | 35 | 47·21 | 1·384 | 35 | 48·44 | ·8847 | 65 | 57·51 |
| 1·288 | 30 | 38·64 | 1·332 | 30 | 39·96 | ·8963 | 60 | 53·78 |
| 1·230 | 25 | 30·75 | 1·279 | 25 | 31·97 | ·9077 | 55 | 49·92 |
| 1·177 | 20 | 23·50 | 1·225 | 20 | 24·50 | ·9188 | 50 | 45·94 |
| 1·128 | 15 | 16·86 | 1·170 | 15 | 17·55 | ·9200* | 49·24 | 45·30 |
| 1·083 | 10 | 10·77 | 1·115 | 10 | 11·15 | ·9296 | 45 | 41·83 |
| 1·041 | 5 | 5·18 | 1·059 | 5 | 5·29 | ·9398 | 40 | 37·59 |
| | | | | | | ·9493 | 35 | 33·23 |
| | | | | | | ·9578 | 30 | 28·73 |

NH$_3$ at 14° C. in

| S.G. | 100 gm. | 100 c.c. |
|---|---|---|
| ·8844 | 36 | 31·84 |
| ·8885 | 34 | 30·21 |
| ·8929 | 32 | 28·57 |
| ·8976 | 30 | 26·93 |
| ·9026 | 28 | 25·27 |
| ·9078 | 26 | 23·60 |
| ·9133 | 24 | 21·92 |
| ·9191 | 22 | 20·22 |
| ·9251 | 20 | 18·50 |
| ·9314 | 18 | 16·77 |
| ·9380 | 16 | 14·91 |
| ·9449 | 14 | 13·23 |
| ·9520 | 12 | 11·42 |
| ·9593 | 10 | 9·59 |

(Alcohol in, continued)

| ·9650 | 25 | 24·12 |
|---|---|---|
| ·9718 | 20 | 19·44 |
| ·9775 | 15 | 14·66 |
| ·9840 | 10 | 9·84 |
| ·9912 | 5 | 4·96 |

To obtain % alcohol by volume multiply the numbers in the last column by 1·2583.

\* "proof spirit."

(Water at 15° C. = 1.)

## V.　Pressure of Aqueous Vapour in mm. of mercury.

| t° C. | mm. | t° C. | mm. | t° C. | mm. | t° C. | Atmos. |
|---|---|---|---|---|---|---|---|
| − 10 | 2·08 | 16 | 13·54 | 90 | 525·39 | 100 | 1·0 |
| − 9 | 2·26 | 17 | 14·42 | 95 | 633·69 | 110 | 1·4 |
| − 8 | 2·46 | 18 | 15·36 | 99 | 733·21 | 120 | 1·96 |
| − 7 | 2·67 | 19 | 16·35 | 99·1 | 735·85 | 130 | 2·67 |
| − 6 | 2·89 | 20 | 17·39 | 99·2 | 738·50 | 140 | 3·57 |
| − 5 | 3·13 | 21 | 18·50 | 99·3 | 741·16 | 150 | 4·7 |
| − 4 | 3·39 | 22 | 19·66 | 99·4 | 743·83 | 160 | 6·1 |
| − 3 | 3·66 | 23 | 20·89 | 99·5 | 746·50 | 170 | 7·8 |
| − 2 | 3·96 | 24 | 22·18 | 99·6 | 749·18 | 180 | 9·9 |
| − 1 | 4·27 | 25 | 23·55 | 99·7 | 751·87 | 190 | 12·4 |
| 0 | 4·60 | 26 | 24·99 | 99·8 | 754·57 | 200 | 15·4 |
| 1 | 4·94 | 27 | 26·51 | 99·9 | 757·28 | 210 | 18·8 |
| 2 | 5·30 | 28 | 28·10 | 100 | 760·00 | 220 | 22·9 |
| 3 | 5·69 | 29 | 29·78 | 100·1 | 762·73 | 230 | 27·5 |
| 4 | 6·10 | 30 | 31·55 | 100·2 | 765·46 | | |
| 5 | 6·53 | 35 | 41·83 | 100·3 | 768·20 | | |
| 6 | 7·00 | 40 | 54·91 | 100·4 | 771·95 | | |
| 7 | 7·49 | 45 | 71·39 | 100·5 | 773·71 | | |
| 8 | 8·02 | 50 | 91·98 | 100·6 | 776·48 | | |
| 9 | 8·57 | 55 | 117·48 | 100·7 | 779·26 | | |
| 10 | 9·17 | 60 | 148·79 | 100·8 | 782·04 | | |
| 11 | 9·79 | 65 | 186·94 | 100·9 | 784·83 | | |
| 12 | 10·46 | 70 | 233·08 | 101 | 787·59 | | |
| 13 | 11·16 | 75 | 288·50 | 105 | 906·41 | | |
| 14 | 11·91 | 80 | 354·62 | 110 | 1075·37 | | |
| 15 | 12·70 | 85 | 433·00 | | | | |

## VI.  *Volume and Specific gravity of Water at different temperatures* (KOPP).

| Temp. °C. | Volume of water (at 0°=1). | Sp. Gr. of water (at 0°=1). | Volume of water (at 4°=1). | Sp. Gr. of water (at 4°=1). |
|---|---|---|---|---|
| 0 | 1·00000 | 1·000000 | 1·00012 | 0·999877 |
| 1 | 0·99995 | 1·000053 | 1·00007 | 0·999930 |
| 2 | 0·99991 | 1·000092 | 1·00003 | 0·999969 |
| 3 | 0·99989 | 1·000115 | 1·00001 | 0·999992 |
| 4 | 0·99988 | 1·000123 | 1·00000 | 1·000000 |
| 5 | 0·99988 | 1·000117 | 1·00001 | 0·999994 |
| 6 | 0·99990 | 1·000097 | 1·00003 | 0·999973 |
| 7 | 0·99994 | 1·000062 | 1·00006 | 0·999939 |
| 8 | 0·99999 | 1·000014 | 1·00011 | 0·999890 |
| 9 | 1·00005 | 0·999952 | 1·00017 | 0·999829 |
| 10 | 1·00012 | 0·999876 | 1·00025 | 0·999753 |
| 11 | 1·00021 | 0·999785 | 1·00034 | 0·999664 |
| 12 | 1·00031 | 0·999686 | 1·00044 | 0·999562 |
| 13 | 1·00043 | 0·999572 | 1·00055 | 0·999449 |
| 14 | 1·00056 | 0·999445 | 1·00068 | 0·999332 |
| 15 | 1·00070 | 0·999306 | 1·00082 | 0·999183 |
| 16 | 1·00085 | 0·999155 | 1·00097 | 0·999032 |
| 17 | 1·00101 | 0·998992 | 1·00113 | 0·998869 |
| 18 | 1·00118 | 0·998817 | 1·00131 | 0·998695 |
| 19 | 1·00137 | 0·998631 | 1·00149 | 0·998509 |
| 20 | 1·00157 | 0·998435 | 1·00169 | 0·998312 |
| 21 | 1·00178 | 0·998228 | 1·00190 | 0·998104 |
| 22 | 1·00200 | 0·998010 | 1·00212 | 0·997886 |
| 23 | 1·00223 | 0·997780 | 1·00235 | 0·997657 |
| 24 | 1·00247 | 0·997541 | 1·00259 | 0·997419 |
| 25 | 1·00271 | 0·997293 | 1·00284 | 0·997170 |
| 26 | 1·00295 | 0·997035 | 1·00310 | 0·996912 |
| 27 | 1·00319 | 0·996767 | 1·00337 | 0·996644 |
| 28 | 1·00347 | 0·996489 | 1·00365 | 0·996367 |
| 29 | 1·00376 | 0·996202 | 1·00393 | 0·996082 |
| 30 | 1·00406 | 0·995908 | 1·00423 | 0·995787 |
| 40 | 1·00753 | ... | ... | ... |
| 50 | 1·01177 | ... | ... | ... |
| 60 | 1·01695 | ... | ... | ... |
| 70 | 1·02225 | ... | ... | ... |
| 80 | 1·02858 | ... | ... | ... |
| 90 | 1·03540 | ... | ... | ... |
| 100 | 1·04299 | ... | ... | ... |

## VII.   *For the calculation of* $1 + \cdot 00367t$.

If $V$ is the volume, and $d$ is the specific gravity, of a gas at $t^{\circ}$ and $b$ mm. mercury, then the volume at $0^{\circ}$ and 760 mm. $= V^{\circ}$, and the specific gravity $= d^{\circ}$; and

$$V^{\circ} = \frac{V}{1 + \cdot 00367t} \times \frac{b}{760}; \ d^{\circ} = d\,(1 + \cdot 00367t) \times \frac{760}{b}.$$

| $t^{\circ}$ | Num. | Log. | $t^{\circ}$ | Num. | Log. | $t^{\circ}$ | Num. | Log. |
|---|---|---|---|---|---|---|---|---|
| $-2 \cdot 0$ | 0·99268 | 9·99681 | $+12 \cdot 5$ | 1·04575 | 0·01943 | $+26 \cdot 5$ | 1·09699 | 0·04021 |
| 1·5 | 0·99451 | 9·99761 | 13·0 | 1·04758 | 0·02019 | 27·0 | 1·09882 | 0·04093 |
| 1·0 | 0·99634 | 9·99841 | 13·5 | 1·04941 | 0·02095 | 27·5 | 1·10065 | 0·04165 |
| 0·5 | 0·99817 | 9·99920 | 14·0 | 1·05124 | 0·02170 | 28·0 | 1·10248 | 0·04237 |
| 0·0 | 1·00000 | 0·00000 | 14·5 | 1·05307 | 0·02246 | 28·5 | 1·10431 | 0·04309 |
| $+0 \cdot 5$ | 1·00183 | 0·00079 | 15·0 | 1·05490 | 0·02321 | 29·0 | 1·10614 | 0·04381 |
| 1·0 | 1·00366 | 0·00159 | 15·5 | 1·05673 | 0·02396 | 29·5 | 1·10797 | 0·04453 |
| 1·5 | 1·00549 | 0·00238 | 16·0 | 1·05856 | 0·02471 | 30·0 | 1·10980 | 0·04524 |
| 2·0 | 1·00732 | 0·00347 | 16·5 | 1·06039 | 0·02546 | 30·5 | 1·11163 | 0·04595 |
| 2·5 | 1·00915 | 0·00397 | 17·0 | 1·06222 | 0·02621 | 31·0 | 1·11346 | 0·04667 |
| 3·0 | 1·01098 | 0·00474 | 17·5 | 1·06405 | 0·02696 | 31·5 | 1·11529 | 0·04738 |
| 3·5 | 1·01281 | 0·00553 | 18·0 | 1·06588 | 0·02771 | 32·0 | 1·11712 | 0·04810 |
| 4·0 | 1·01464 | 0·00631 | 18·5 | 1·06771 | 0·02846 | 32·5 | 1·11895 | 0·04881 |
| 4·5 | 1·01647 | 0·00710 | 19·0 | 1·06954 | 0·02921 | 33·0 | 1·12078 | 0·04952 |
| 5·0 | 1·01830 | 0·00788 | 19·5 | 1·07137 | 0·02995 | 33·5 | 1·12261 | 0·05022 |
| 5·5 | 1·02013 | 0·00865 | 20·0 | 1·07320 | 0·03068 | 34·0 | 1·12444 | 0·05094 |
| 6·0 | 1·02196 | 0·00943 | 20·5 | 1·07503 | 0·03142 | 34·5 | 1·12627 | 0·05164 |
| 6·5 | 1·02379 | 0·01022 | 21·0 | 1·07686 | 0·03216 | 35·0 | 1·12810 | 0·05235 |
| 7·0 | 1·02562 | 0·01099 | 21·5 | 1·07869 | 0·03290 | 35·5 | 1·12993 | 0·05305 |
| 7·5 | 1·02745 | 0·01177 | 22·0 | 1·08052 | 0·03363 | 36·0 | 1·13176 | 0·05375 |
| 8·0 | 1·02928 | 0·01253 | 22·5 | 1·08235 | 0·03437 | 36·5 | 1·13359 | 0·05446 |
| 8·5 | 1·03111 | 0·01330 | 23·0 | 1·08418 | 0·03510 | 37·0 | 1·13542 | 0·05516 |
| 9·0 | 1·03294 | 0·01407 | 23·5 | 1·08601 | 0·03583 | 37·5 | 1·13725 | 0·05585 |
| 9·5 | 1·03477 | 0·01484 | 24·0 | 1·08784 | 0·03656 | 38·0 | 1·13908 | 0·05655 |
| 10·0 | 1·03660 | 0·01561 | 24·5 | 1·08967 | 0·03729 | 38·5 | 1·14901 | 0·05725 |
| 10·5 | 1·03843 | 0·01639 | 25·0 | 1·09150 | 0·03802 | 39·0 | 1·14274 | 0·05795 |
| 11·0 | 1·04026 | 0·01714 | 25·5 | 1·09333 | 0·03875 | 39·5 | 1·14457 | 0·05864 |
| 11·5 | 1·04209 | 0·01790 | 26·0 | 1·09516 | 0·03948 | 40·0 | 1·14640 | 0·05933 |
| 12·0 | 1·04392 | 0·01867 | | | | | | |

14—2

VIII. *For the calculation of* $\log \dfrac{\cdot001293}{(1 + \cdot00367t)\,760} = \log a.$

| $t°$ | log. $a$ | $t°$ | log. $a$ | $t°$ | log. $a$ | $t°$ | log. $a$ | $t°$ | log. $a$ |
|---|---|---|---|---|---|---|---|---|---|
| −5 | 6·23889 | 5 | 6·222955 | 15 | 6·207579 | 25 | 6·192728 | 35 | 6·178369 |
| −4 | 23727 | 6 | 221392 | 16 | 206071 | 26 | 191271 | 36 | 176959 |
| −3 | 23566 | 7 | 219835 | 17 | 204568 | 27 | 189818 | 37 | 175554 |
| −2 | 23405 | 8 | 218284 | 18 | 203070 | 28 | 188371 | 38 | 174153 |
| −1 | 23245 | 9 | 216739 | 19 | 201577 | 29 | 186928 | 39 | 172756 |
| 0 | 230852 | 10 | 215199 | 20 | 200090 | 30 | 185490 | 40 | 171364 |
| +1 | 229261 | 11 | 213664 | 21 | 198608 | 31 | 184056 | 41 | 169976 |
| 2 | 227666 | 12 | 212135 | 22 | 197130 | 32 | 182927 | 42 | 168593 |
| 3 | 226069 | 13 | 210611 | 23 | 195658 | 33 | 181203 | 43 | 167214 |
| 4 | 224523 | 14 | 209092 | 24 | 194191 | 34 | 179784 | 44 | 165840 |

IX. *Weight in grams of 1 litre of various gases at 0° and 760 mm.*

| Gas. | Weight in grams of 1 litre. | Gas. | Weight in grams of 1 litre. |
|---|---|---|---|
| O | 1·430 | $NO_2$ | 2·06 |
| H | 0·08958 | $SO_2$ | 2·87 |
| N | 1·256 | CO | 1·254 |
| Cl | 3·18 | $CO_2$ | 1·9774 |
| Br | 7·16 | $Cl_2O$ | 3·90 |
| I | 11·3 | $COCl_2$ | 4·43 |
| Hg | 8·9 | $CH_3Cl$ | 2·261 |
| HCl | 1·635 | $C_2H_5Cl$ | 2·889 |
| HBr | 3·63 | $BCl_3$ | 5·26 |
| HI | 5·73 | $BF_3$ | 3·05 |
| HF | 0·896 | $SiF_4$ | 4·66 |
| $H_2S$ | 1·523 | $CH_4$ | 0·716 |
| $H_2Se$ | 3·63 | $C_2H_6$ | 1·343 |
| $H_3N$ | 0·761 | $C_2H_4$ | 1·254 |
| $H_3P$ | 1·523 | $C_2H_2$ | 1·165 |
| $H_3As$ | 3·49 | $C_2N_2$ | 2·330 |
| $N_2O$ | 1·977 | CNH | 1·210 |
| NO | 1·343 | CNCl | 2·755 |
| $N_2O_4$ | 4·12 | $H_2O$ | 0·806 |

## LOGARITHMS OF NUMBERS.

| Natural Numbers. | 0 | 1 | 2 | 3 | 4 | 5 | 6 | 7 | 8 | 9 | Proportional Parts. | | | | | | | | |
|---|---|---|---|---|---|---|---|---|---|---|---|---|---|---|---|---|---|---|---|
| | | | | | | | | | | | 1 | 2 | 3 | 4 | 5 | 6 | 7 | 8 | 9 |
| 10 | 0000 | 0043 | 0086 | 0128 | 0170 | 0212 | 0253 | 0294 | 0334 | 0374 | 4 | 8 | 12 | 17 | 21 | 25 | 29 | 33 | 37 |
| 11 | 0414 | 0453 | 0492 | 0531 | 0569 | 0607 | 0645 | 0682 | 0719 | 0755 | 4 | 8 | 11 | 15 | 19 | 23 | 26 | 30 | 34 |
| 12 | 0792 | 0828 | 0864 | 0899 | 0934 | 0969 | 1004 | 1038 | 1072 | 1106 | 3 | 7 | 10 | 14 | 17 | 21 | 24 | 28 | 31 |
| 13 | 1139 | 1173 | 1206 | 1239 | 1271 | 1303 | 1335 | 1367 | 1399 | 1430 | 3 | 6 | 10 | 13 | 16 | 19 | 23 | 26 | 29 |
| 14 | 1461 | 1492 | 1523 | 1553 | 1584 | 1614 | 1644 | 1673 | 1703 | 1732 | 3 | 6 | 9 | 12 | 15 | 18 | 21 | 24 | 27 |
| 15 | 1761 | 1790 | 1818 | 1847 | 1875 | 1903 | 1931 | 1959 | 1987 | 2014 | 3 | 6 | 8 | 11 | 14 | 17 | 20 | 22 | 25 |
| 16 | 2041 | 2068 | 2095 | 2122 | 2148 | 2175 | 2201 | 2227 | 2253 | 2279 | 3 | 5 | 8 | 11 | 13 | 16 | 18 | 21 | 24 |
| 17 | 2304 | 2330 | 2355 | 2380 | 2405 | 2430 | 2455 | 2480 | 2504 | 2529 | 2 | 5 | 7 | 10 | 12 | 15 | 17 | 20 | 22 |
| 18 | 2553 | 2577 | 2601 | 2625 | 2648 | 2672 | 2695 | 2718 | 2742 | 2765 | 2 | 5 | 7 | 9 | 12 | 14 | 16 | 19 | 21 |
| 19 | 2788 | 2810 | 2833 | 2856 | 2878 | 2900 | 2923 | 2945 | 2967 | 2989 | 2 | 4 | 7 | 9 | 11 | 13 | 16 | 18 | 20 |
| 20 | 3010 | 3032 | 3054 | 3075 | 3096 | 3118 | 3139 | 3160 | 3181 | 3201 | 2 | 4 | 6 | 8 | 11 | 13 | 15 | 17 | 19 |
| 21 | 3222 | 3243 | 3263 | 3284 | 3304 | 3324 | 3345 | 3365 | 3385 | 3404 | 2 | 4 | 6 | 8 | 10 | 12 | 14 | 16 | 18 |
| 22 | 3424 | 3444 | 3464 | 3483 | 3502 | 3522 | 3541 | 3560 | 3579 | 3598 | 2 | 4 | 6 | 8 | 10 | 12 | 14 | 15 | 17 |
| 23 | 3617 | 3636 | 3655 | 3674 | 3692 | 3711 | 3729 | 3747 | 3766 | 3784 | 2 | 4 | 6 | 7 | 9 | 11 | 13 | 15 | 17 |
| 24 | 3802 | 3820 | 3838 | 3856 | 3874 | 3892 | 3909 | 3927 | 3945 | 3962 | 2 | 4 | 5 | 7 | 9 | 11 | 12 | 14 | 16 |
| 25 | 3979 | 3997 | 4014 | 4031 | 4048 | 4065 | 4082 | 4099 | 4116 | 4133 | 2 | 3 | 5 | 7 | 9 | 10 | 12 | 14 | 15 |
| 26 | 4150 | 4166 | 4183 | 4200 | 4216 | 4232 | 4249 | 4265 | 4281 | 4298 | 2 | 3 | 5 | 7 | 8 | 10 | 11 | 13 | 15 |
| 27 | 4314 | 4330 | 4346 | 4362 | 4378 | 4393 | 4409 | 4425 | 4440 | 4456 | 2 | 3 | 5 | 6 | 8 | 9 | 11 | 13 | 14 |
| 28 | 4472 | 4487 | 4502 | 4518 | 4533 | 4548 | 4564 | 4579 | 4594 | 4609 | 2 | 3 | 5 | 6 | 8 | 9 | 11 | 12 | 14 |
| 29 | 4624 | 4639 | 4654 | 4669 | 4683 | 4698 | 4713 | 4728 | 4742 | 4757 | 1 | 3 | 4 | 6 | 7 | 9 | 10 | 12 | 13 |
| 30 | 4771 | 4786 | 4800 | 4814 | 4829 | 4843 | 4857 | 4871 | 4886 | 4900 | 1 | 3 | 4 | 6 | 7 | 9 | 10 | 11 | 13 |
| 31 | 4914 | 4928 | 4942 | 4955 | 4969 | 4983 | 4997 | 5011 | 5024 | 5038 | 1 | 3 | 4 | 6 | 7 | 8 | 10 | 11 | 12 |
| 32 | 5051 | 5065 | 5079 | 5092 | 5105 | 5119 | 5132 | 5145 | 5159 | 5172 | 1 | 3 | 4 | 5 | 7 | 8 | 9 | 11 | 12 |
| 33 | 5185 | 5198 | 5211 | 5224 | 5237 | 5250 | 5263 | 5276 | 5289 | 5302 | 1 | 3 | 4 | 5 | 6 | 8 | 9 | 10 | 12 |
| 34 | 5315 | 5328 | 5340 | 5353 | 5366 | 5378 | 5391 | 5403 | 5416 | 5428 | 1 | 3 | 4 | 5 | 6 | 8 | 9 | 10 | 11 |
| 35 | 5441 | 5453 | 5465 | 5478 | 5490 | 5502 | 5514 | 5527 | 5539 | 5551 | 1 | 2 | 4 | 5 | 6 | 7 | 9 | 10 | 11 |
| 36 | 5563 | 5575 | 5587 | 5599 | 5611 | 5623 | 5635 | 5647 | 5658 | 5670 | 1 | 2 | 4 | 5 | 6 | 7 | 8 | 10 | 11 |
| 37 | 5682 | 5694 | 5705 | 5717 | 5729 | 5740 | 5752 | 5763 | 5775 | 5786 | 1 | 2 | 3 | 5 | 6 | 7 | 8 | 9 | 10 |
| 38 | 5798 | 5809 | 5821 | 5832 | 5843 | 5855 | 5866 | 5877 | 5888 | 5899 | 1 | 2 | 3 | 5 | 6 | 7 | 8 | 9 | 10 |
| 39 | 5911 | 5922 | 5933 | 5944 | 5955 | 5966 | 5977 | 5988 | 5999 | 6010 | 1 | 2 | 3 | 4 | 5 | 7 | 8 | 9 | 10 |
| 40 | 6021 | 6031 | 6042 | 6053 | 6064 | 6075 | 6085 | 6096 | 6107 | 6117 | 1 | 2 | 3 | 4 | 5 | 6 | 8 | 9 | 10 |
| 41 | 6128 | 6138 | 6149 | 6160 | 6170 | 6180 | 6191 | 6201 | 6212 | 6222 | 1 | 2 | 3 | 4 | 5 | 6 | 7 | 8 | 9 |
| 42 | 6232 | 6243 | 6253 | 6263 | 6274 | 6284 | 6294 | 6304 | 6314 | 6325 | 1 | 2 | 3 | 4 | 5 | 6 | 7 | 8 | 9 |
| 43 | 6335 | 6345 | 6355 | 6365 | 6375 | 6385 | 6395 | 6405 | 6415 | 6425 | 1 | 2 | 3 | 4 | 5 | 6 | 7 | 8 | 9 |
| 44 | 6435 | 6444 | 6454 | 6464 | 6474 | 6484 | 6493 | 6503 | 6513 | 6522 | 1 | 2 | 3 | 4 | 5 | 6 | 7 | 8 | 9 |
| 45 | 6532 | 6542 | 6551 | 6561 | 6571 | 6580 | 6590 | 6599 | 6609 | 6618 | 1 | 2 | 3 | 4 | 5 | 6 | 7 | 8 | 9 |
| 46 | 6628 | 6637 | 6646 | 6656 | 6665 | 6675 | 6684 | 6693 | 6702 | 6712 | 1 | 2 | 3 | 4 | 5 | 6 | 7 | 7 | 8 |
| 47 | 6721 | 6730 | 6739 | 6749 | 6758 | 6767 | 6776 | 6785 | 6794 | 6803 | 1 | 2 | 3 | 4 | 5 | 5 | 6 | 7 | 8 |
| 48 | 6812 | 6821 | 6830 | 6839 | 6848 | 6857 | 6866 | 6875 | 6884 | 6893 | 1 | 2 | 3 | 4 | 4 | 5 | 6 | 7 | 8 |
| 49 | 6902 | 6911 | 6920 | 6928 | 6937 | 6946 | 6955 | 6964 | 6972 | 6981 | 1 | 2 | 3 | 4 | 4 | 5 | 6 | 7 | 8 |
| 50 | 6990 | 6998 | 7007 | 7016 | 7024 | 7033 | 7042 | 7050 | 7059 | 7067 | 1 | 2 | 3 | 3 | 4 | 5 | 6 | 7 | 8 |
| 51 | 7076 | 7084 | 7093 | 7101 | 7110 | 7118 | 7126 | 7135 | 7143 | 7152 | 1 | 2 | 3 | 3 | 4 | 5 | 6 | 7 | 8 |
| 52 | 7160 | 7168 | 7177 | 7185 | 7193 | 7202 | 7210 | 7218 | 7226 | 7235 | 1 | 2 | 2 | 3 | 4 | 5 | 6 | 7 | 7 |
| 53 | 7243 | 7251 | 7259 | 7267 | 7275 | 7284 | 7292 | 7300 | 7308 | 7316 | 1 | 2 | 2 | 3 | 4 | 5 | 6 | 6 | 7 |
| 54 | 7324 | 7332 | 7340 | 7348 | 7356 | 7364 | 7372 | 7380 | 7388 | 7396 | 1 | 2 | 2 | 3 | 4 | 5 | 6 | 6 | 7 |

## LOGARITHMS OF NUMBERS.

| Natural Numbers | 0 | 1 | 2 | 3 | 4 | 5 | 6 | 7 | 8 | 9 | Proportional Parts. 1 | 2 | 3 | 4 | 5 | 6 | 7 | 8 | 9 |
|---|---|---|---|---|---|---|---|---|---|---|---|---|---|---|---|---|---|---|---|
| 55 | 7404 | 7412 | 7419 | 7427 | 7435 | 7443 | 7451 | 7459 | 7466 | 7474 | 1 | 2 | 2 | 3 | 4 | 5 | 5 | 6 | 7 |
| 56 | 7482 | 7490 | 7497 | 7505 | 7513 | 7520 | 7528 | 7536 | 7543 | 7551 | 1 | 2 | 2 | 3 | 4 | 5 | 5 | 6 | 7 |
| 57 | 7559 | 7566 | 7574 | 7582 | 7589 | 7597 | 7604 | 7612 | 7619 | 7627 | 1 | 2 | 2 | 3 | 4 | 5 | 5 | 6 | 7 |
| 58 | 7634 | 7642 | 7649 | 7657 | 7664 | 7672 | 7679 | 7686 | 7694 | 7701 | 1 | 1 | 2 | 3 | 4 | 4 | 5 | 6 | 7 |
| 59 | 7709 | 7716 | 7723 | 7731 | 7738 | 7745 | 7752 | 7760 | 7767 | 7774 | 1 | 1 | 2 | 3 | 4 | 4 | 5 | 6 | 7 |
| 60 | 7782 | 7789 | 7796 | 7803 | 7810 | 7818 | 7825 | 7832 | 7839 | 7846 | 1 | 1 | 2 | 3 | 4 | 4 | 5 | 6 | 6 |
| 61 | 7853 | 7860 | 7868 | 7875 | 7882 | 7889 | 7896 | 7903 | 7910 | 7917 | 1 | 1 | 2 | 3 | 4 | 4 | 5 | 6 | 6 |
| 62 | 7924 | 7931 | 7938 | 7945 | 7952 | 7959 | 7966 | 7973 | 7980 | 7987 | 1 | 1 | 2 | 3 | 3 | 4 | 5 | 6 | 6 |
| 63 | 7993 | 8000 | 8007 | 8014 | 8021 | 8028 | 8035 | 8041 | 8048 | 8055 | 1 | 1 | 2 | 3 | 3 | 4 | 5 | 5 | 6 |
| 64 | 8062 | 8069 | 8075 | 8082 | 8089 | 8096 | 8102 | 8109 | 8116 | 8122 | 1 | 1 | 2 | 3 | 3 | 4 | 5 | 5 | 6 |
| 65 | 8129 | 8136 | 8142 | 8149 | 8156 | 8162 | 8169 | 8176 | 8182 | 8189 | 1 | 1 | 2 | 3 | 3 | 4 | 5 | 5 | 6 |
| 66 | 8195 | 8202 | 8209 | 8215 | 8222 | 8228 | 8235 | 8241 | 8248 | 8254 | 1 | 1 | 2 | 3 | 3 | 4 | 5 | 5 | 6 |
| 67 | 8261 | 8267 | 8274 | 8280 | 8287 | 8293 | 8299 | 8306 | 8312 | 8319 | 1 | 1 | 2 | 3 | 3 | 4 | 5 | 5 | 6 |
| 68 | 8325 | 8331 | 8338 | 8344 | 8351 | 8357 | 8363 | 8370 | 8376 | 8382 | 1 | 1 | 2 | 3 | 3 | 4 | 4 | 5 | 6 |
| 69 | 8388 | 8395 | 8401 | 8407 | 8414 | 8420 | 8426 | 8432 | 8439 | 8445 | 1 | 1 | 2 | 2 | 3 | 4 | 4 | 5 | 6 |
| 70 | 8451 | 8457 | 8463 | 8470 | 8476 | 8482 | 8488 | 8494 | 8500 | 8506 | 1 | 1 | 2 | 2 | 3 | 4 | 4 | 5 | 6 |
| 71 | 8513 | 8519 | 8525 | 8531 | 8537 | 8543 | 8549 | 8555 | 8561 | 8567 | 1 | 1 | 2 | 2 | 3 | 4 | 4 | 5 | 5 |
| 72 | 8573 | 8579 | 8585 | 8591 | 8597 | 8603 | 8609 | 8615 | 8621 | 8627 | 1 | 1 | 2 | 2 | 3 | 4 | 4 | 5 | 5 |
| 73 | 8633 | 8639 | 8645 | 8651 | 8657 | 8663 | 8669 | 8675 | 8681 | 8686 | 1 | 1 | 2 | 2 | 3 | 4 | 4 | 5 | 5 |
| 74 | 8692 | 8698 | 8704 | 8710 | 8716 | 8722 | 8727 | 8733 | 8739 | 8745 | 1 | 1 | 2 | 2 | 3 | 4 | 4 | 5 | 5 |
| 75 | 8751 | 8756 | 8762 | 8768 | 8774 | 8779 | 8785 | 8791 | 8797 | 8802 | 1 | 1 | 2 | 2 | 3 | 3 | 4 | 5 | 5 |
| 76 | 8808 | 8814 | 8820 | 8825 | 8831 | 8837 | 8842 | 8848 | 8854 | 8859 | 1 | 1 | 2 | 2 | 3 | 3 | 4 | 5 | 5 |
| 77 | 8865 | 8871 | 8876 | 8882 | 8887 | 8893 | 8899 | 8904 | 8910 | 8915 | 1 | 1 | 2 | 2 | 3 | 3 | 4 | 4 | 5 |
| 78 | 8921 | 8927 | 8932 | 8938 | 8943 | 8949 | 8954 | 8960 | 8965 | 8971 | 1 | 1 | 2 | 2 | 3 | 3 | 4 | 4 | 5 |
| 79 | 8976 | 8982 | 8987 | 8993 | 8998 | 9004 | 9009 | 9015 | 9020 | 9025 | 1 | 1 | 2 | 2 | 3 | 3 | 4 | 4 | 5 |
| 80 | 9031 | 9036 | 9042 | 9047 | 9053 | 9058 | 9063 | 9069 | 9074 | 9079 | 1 | 1 | 2 | 2 | 3 | 3 | 4 | 4 | 5 |
| 81 | 9085 | 9090 | 9096 | 9101 | 9106 | 9112 | 9117 | 9122 | 9128 | 9133 | 1 | 1 | 2 | 2 | 3 | 3 | 4 | 4 | 5 |
| 82 | 9138 | 9143 | 9149 | 9154 | 9159 | 9165 | 9170 | 9175 | 9180 | 9186 | 1 | 1 | 2 | 2 | 3 | 3 | 4 | 4 | 5 |
| 83 | 9191 | 9196 | 9201 | 9206 | 9212 | 9217 | 9222 | 9227 | 9232 | 9238 | 1 | 1 | 2 | 2 | 3 | 3 | 4 | 4 | 5 |
| 84 | 9243 | 9248 | 9253 | 9258 | 9263 | 9269 | 9274 | 9279 | 9284 | 9289 | 1 | 1 | 2 | 2 | 3 | 3 | 4 | 4 | 5 |
| 85 | 9294 | 9299 | 9304 | 9309 | 9315 | 9320 | 9325 | 9330 | 9335 | 9340 | 1 | 1 | 2 | 2 | 3 | 3 | 4 | 4 | 5 |
| 86 | 9345 | 9350 | 9355 | 9360 | 9365 | 9370 | 9375 | 9380 | 9385 | 9390 | 1 | 1 | 2 | 2 | 3 | 3 | 4 | 4 | 5 |
| 87 | 9395 | 9400 | 9405 | 9410 | 9415 | 9420 | 9425 | 9430 | 9435 | 9440 | 0 | 1 | 1 | 2 | 2 | 3 | 3 | 4 | 4 |
| 88 | 9445 | 9450 | 9455 | 9460 | 9465 | 9469 | 9474 | 9479 | 9484 | 9489 | 0 | 1 | 1 | 2 | 2 | 3 | 3 | 4 | 4 |
| 89 | 9494 | 9499 | 9504 | 9509 | 9513 | 9518 | 9523 | 9528 | 9533 | 9538 | 0 | 1 | 1 | 2 | 2 | 3 | 3 | 4 | 4 |
| 90 | 9542 | 9547 | 9552 | 9557 | 9562 | 9566 | 9571 | 9576 | 9581 | 9586 | 0 | 1 | 1 | 2 | 2 | 3 | 3 | 4 | 4 |
| 91 | 9590 | 9595 | 9600 | 9605 | 9609 | 9614 | 9619 | 9624 | 9628 | 9633 | 0 | 1 | 1 | 2 | 2 | 3 | 3 | 4 | 4 |
| 92 | 9638 | 9643 | 9647 | 9652 | 9657 | 9661 | 9666 | 9671 | 9675 | 9680 | 0 | 1 | 1 | 2 | 2 | 3 | 3 | 4 | 4 |
| 93 | 9685 | 9689 | 9694 | 9699 | 9703 | 9708 | 9713 | 9717 | 9722 | 9727 | 0 | 1 | 1 | 2 | 2 | 3 | 3 | 4 | 4 |
| 94 | 9731 | 9736 | 9741 | 9745 | 9750 | 9754 | 9759 | 9763 | 9768 | 9773 | 0 | 1 | 1 | 2 | 2 | 3 | 3 | 4 | 4 |
| 95 | 9777 | 9782 | 9786 | 9791 | 9795 | 9800 | 9805 | 9809 | 9814 | 9818 | 0 | 1 | 1 | 2 | 2 | 3 | 3 | 4 | 4 |
| 96 | 9823 | 9827 | 9832 | 9836 | 9841 | 9845 | 9850 | 9854 | 9859 | 9863 | 0 | 1 | 1 | 2 | 2 | 3 | 3 | 4 | 4 |
| 97 | 9868 | 9872 | 9877 | 9881 | 9886 | 9890 | 9894 | 9899 | 9903 | 9908 | 0 | 1 | 1 | 2 | 2 | 3 | 3 | 4 | 4 |
| 98 | 9912 | 9917 | 9921 | 9926 | 9930 | 9934 | 9939 | 9943 | 9948 | 9952 | 0 | 1 | 1 | 2 | 2 | 3 | 3 | 4 | 4 |
| 99 | 9956 | 9961 | 9965 | 9969 | 9974 | 9978 | 9983 | 9987 | 9991 | 9996 | 0 | 1 | 1 | 2 | 2 | 3 | 3 | 3 | 4 |

## ANTILOGARITHMS.

| Logarithms | 0 | 1 | 2 | 3 | 4 | 5 | 6 | 7 | 8 | 9 | Proportional Parts. | | | | | | | | |
|---|---|---|---|---|---|---|---|---|---|---|---|---|---|---|---|---|---|---|---|
| | | | | | | | | | | | 1 | 2 | 3 | 4 | 5 | 6 | 7 | 8 | 9 |
| ·00 | 1000 | 1002 | 1005 | 1007 | 1009 | 1012 | 1014 | 1016 | 1019 | 1021 | 0 | 0 | 1 | 1 | 1 | 1 | 2 | 2 | 2 |
| ·01 | 1023 | 1026 | 1028 | 1030 | 1033 | 1035 | 1038 | 1040 | 1042 | 1045 | 0 | 0 | 1 | 1 | 1 | 1 | 2 | 2 | 2 |
| ·02 | 1047 | 1050 | 1052 | 1054 | 1057 | 1059 | 1062 | 1064 | 1067 | 1069 | 0 | 0 | 1 | 1 | 1 | 1 | 2 | 2 | 2 |
| ·03 | 1072 | 1074 | 1076 | 1079 | 1081 | 1084 | 1086 | 1089 | 1091 | 1094 | 0 | 0 | 1 | 1 | 1 | 1 | 2 | 2 | 2 |
| ·04 | 1096 | 1099 | 1102 | 1104 | 1107 | 1109 | 1112 | 1114 | 1117 | 1119 | 0 | 1 | 1 | 1 | 1 | 2 | 2 | 2 | 2 |
| ·05 | 1122 | 1125 | 1127 | 1130 | 1132 | 1135 | 1138 | 1140 | 1143 | 1146 | 0 | 1 | 1 | 1 | 1 | 2 | 2 | 2 | 2 |
| ·06 | 1148 | 1151 | 1153 | 1156 | 1159 | 1161 | 1164 | 1167 | 1169 | 1172 | 0 | 1 | 1 | 1 | 1 | 2 | 2 | 2 | 2 |
| ·07 | 1175 | 1178 | 1180 | 1183 | 1186 | 1189 | 1191 | 1194 | 1197 | 1199 | 0 | 1 | 1 | 1 | 1 | 2 | 2 | 2 | 2 |
| ·08 | 1202 | 1205 | 1208 | 1211 | 1213 | 1216 | 1219 | 1222 | 1225 | 1227 | 0 | 1 | 1 | 1 | 1 | 2 | 2 | 2 | 3 |
| ·09 | 1230 | 1233 | 1236 | 1239 | 1242 | 1245 | 1247 | 1250 | 1253 | 1256 | 0 | 1 | 1 | 1 | 1 | 2 | 2 | 2 | 3 |
| ·10 | 1259 | 1262 | 1265 | 1268 | 1271 | 1274 | 1276 | 1279 | 1282 | 1285 | 0 | 1 | 1 | 1 | 1 | 2 | 2 | 2 | 3 |
| ·11 | 1288 | 1291 | 1294 | 1297 | 1300 | 1303 | 1306 | 1309 | 1312 | 1315 | 0 | 1 | 1 | 1 | 2 | 2 | 2 | 2 | 3 |
| ·12 | 1318 | 1321 | 1324 | 1327 | 1330 | 1334 | 1337 | 1340 | 1343 | 1346 | 0 | 1 | 1 | 1 | 2 | 2 | 2 | 2 | 3 |
| ·13 | 1349 | 1352 | 1355 | 1358 | 1361 | 1365 | 1368 | 1371 | 1374 | 1377 | 0 | 1 | 1 | 1 | 2 | 2 | 2 | 3 | 3 |
| ·14 | 1380 | 1384 | 1387 | 1390 | 1393 | 1396 | 1400 | 1403 | 1406 | 1409 | 0 | 1 | 1 | 1 | 2 | 2 | 2 | 3 | 3 |
| ·15 | 1413 | 1416 | 1419 | 1422 | 1426 | 1429 | 1432 | 1435 | 1439 | 1442 | 0 | 1 | 1 | 1 | 2 | 2 | 2 | 3 | 3 |
| ·16 | 1445 | 1449 | 1452 | 1455 | 1459 | 1462 | 1466 | 1469 | 1472 | 1476 | 0 | 1 | 1 | 1 | 2 | 2 | 2 | 3 | 3 |
| ·17 | 1479 | 1483 | 1486 | 1489 | 1493 | 1496 | 1500 | 1503 | 1507 | 1510 | 0 | 1 | 1 | 1 | 2 | 2 | 2 | 3 | 3 |
| ·18 | 1514 | 1517 | 1521 | 1524 | 1528 | 1531 | 1535 | 1538 | 1542 | 1545 | 0 | 1 | 1 | 1 | 2 | 2 | 2 | 3 | 3 |
| ·19 | 1549 | 1552 | 1556 | 1560 | 1563 | 1567 | 1570 | 1574 | 1578 | 1581 | 0 | 1 | 1 | 1 | 2 | 2 | 3 | 3 | 3 |
| ·20 | 1585 | 1589 | 1592 | 1596 | 1600 | 1603 | 1607 | 1611 | 1614 | 1618 | 0 | 1 | 1 | 1 | 2 | 2 | 3 | 3 | 3 |
| ·21 | 1622 | 1626 | 1629 | 1633 | 1637 | 1641 | 1644 | 1648 | 1652 | 1656 | 0 | 1 | 1 | 2 | 2 | 2 | 3 | 3 | 3 |
| ·22 | 1660 | 1663 | 1667 | 1671 | 1675 | 1679 | 1683 | 1687 | 1690 | 1694 | 0 | 1 | 1 | 2 | 2 | 2 | 3 | 3 | 3 |
| ·23 | 1698 | 1702 | 1706 | 1710 | 1714 | 1718 | 1722 | 1726 | 1730 | 1734 | 0 | 1 | 1 | 2 | 2 | 2 | 3 | 3 | 4 |
| ·24 | 1738 | 1742 | 1746 | 1750 | 1754 | 1758 | 1762 | 1766 | 1770 | 1774 | 0 | 1 | 1 | 2 | 2 | 2 | 3 | 3 | 4 |
| ·25 | 1778 | 1782 | 1786 | 1791 | 1795 | 1799 | 1803 | 1807 | 1811 | 1816 | 0 | 1 | 1 | 2 | 2 | 2 | 3 | 3 | 4 |
| ·26 | 1820 | 1824 | 1828 | 1832 | 1837 | 1841 | 1845 | 1849 | 1854 | 1858 | 0 | 1 | 1 | 2 | 2 | 3 | 3 | 3 | 4 |
| ·27 | 1862 | 1866 | 1871 | 1875 | 1879 | 1884 | 1888 | 1892 | 1897 | 1901 | 0 | 1 | 1 | 2 | 2 | 3 | 3 | 4 | 4 |
| ·28 | 1905 | 1910 | 1914 | 1919 | 1923 | 1928 | 1932 | 1936 | 1941 | 1945 | 0 | 1 | 1 | 2 | 2 | 3 | 3 | 4 | 4 |
| ·29 | 1950 | 1954 | 1959 | 1963 | 1968 | 1972 | 1977 | 1982 | 1986 | 1991 | 0 | 1 | 1 | 2 | 2 | 3 | 3 | 4 | 4 |
| ·30 | 1995 | 2000 | 2004 | 2009 | 2014 | 2018 | 2023 | 2028 | 2032 | 2037 | 0 | 1 | 1 | 2 | 2 | 3 | 3 | 4 | 4 |
| ·31 | 2042 | 2046 | 2051 | 2056 | 2061 | 2065 | 2070 | 2075 | 2080 | 2084 | 0 | 1 | 1 | 2 | 2 | 3 | 3 | 4 | 4 |
| ·32 | 2089 | 2094 | 2099 | 2104 | 2109 | 2113 | 2118 | 2123 | 2128 | 2133 | 0 | 1 | 1 | 2 | 2 | 3 | 3 | 4 | 4 |
| ·33 | 2138 | 2143 | 2148 | 2153 | 2158 | 2163 | 2168 | 2173 | 2178 | 2183 | 0 | 1 | 1 | 2 | 2 | 3 | 3 | 4 | 4 |
| ·34 | 2188 | 2193 | 2198 | 2203 | 2208 | 2213 | 2218 | 2223 | 2228 | 2234 | 1 | 1 | 2 | 2 | 3 | 3 | 4 | 4 | 5 |
| ·35 | 2239 | 2244 | 2249 | 2254 | 2259 | 2265 | 2270 | 2275 | 2280 | 2286 | 1 | 1 | 2 | 2 | 3 | 3 | 4 | 4 | 5 |
| ·36 | 2291 | 2296 | 2301 | 2307 | 2312 | 2317 | 2323 | 2328 | 2333 | 2339 | 1 | 1 | 2 | 2 | 3 | 3 | 4 | 4 | 5 |
| ·37 | 2344 | 2350 | 2355 | 2360 | 2366 | 2371 | 2377 | 2382 | 2388 | 2393 | 1 | 1 | 2 | 2 | 3 | 3 | 4 | 4 | 5 |
| ·38 | 2399 | 2404 | 2410 | 2415 | 2421 | 2427 | 2432 | 2438 | 2443 | 2449 | 1 | 1 | 2 | 2 | 3 | 3 | 4 | 4 | 5 |
| ·39 | 2455 | 2460 | 2466 | 2472 | 2477 | 2483 | 2489 | 2495 | 2500 | 2506 | 1 | 1 | 2 | 2 | 3 | 3 | 4 | 5 | 5 |
| ·40 | 2512 | 2518 | 2523 | 2529 | 2535 | 2541 | 2547 | 2553 | 2559 | 2564 | 1 | 1 | 2 | 2 | 3 | 4 | 4 | 5 | 5 |
| ·41 | 2570 | 2576 | 2582 | 2588 | 2594 | 2600 | 2606 | 2612 | 2618 | 2624 | 1 | 1 | 2 | 2 | 3 | 4 | 4 | 5 | 5 |
| ·42 | 2630 | 2636 | 2642 | 2649 | 2655 | 2661 | 2667 | 2673 | 2679 | 2685 | 1 | 1 | 2 | 2 | 3 | 4 | 4 | 4 | 6 |
| ·43 | 2692 | 2698 | 2704 | 2710 | 2716 | 2723 | 2729 | 2735 | 2742 | 2748 | 1 | 1 | 2 | 3 | 3 | 4 | 4 | 5 | 6 |
| ·44 | 2754 | 2761 | 2767 | 2773 | 2780 | 2786 | 2793 | 2799 | 2805 | 2812 | 1 | 1 | 2 | 3 | 3 | 4 | 4 | 5 | 6 |
| ·45 | 2818 | 2825 | 2831 | 2838 | 2844 | 2851 | 2858 | 2864 | 2871 | 2877 | 1 | 1 | 2 | 3 | 3 | 4 | 5 | 5 | 6 |
| ·46 | 2884 | 2891 | 2897 | 2904 | 2911 | 2917 | 2924 | 2931 | 2938 | 2944 | 1 | 1 | 2 | 3 | 3 | 4 | 5 | 5 | 6 |
| ·47 | 2951 | 2958 | 2965 | 2972 | 2979 | 2985 | 2992 | 2999 | 3006 | 3013 | 1 | 1 | 2 | 3 | 3 | 4 | 5 | 5 | 6 |
| ·48 | 3020 | 3027 | 3034 | 3041 | 3048 | 3055 | 3062 | 3069 | 3076 | 3083 | 1 | 1 | 2 | 3 | 4 | 4 | 5 | 6 | 6 |
| ·49 | 3090 | 3097 | 3105 | 3112 | 3119 | 3126 | 3133 | 3141 | 3148 | 3155 | 1 | 1 | 2 | 3 | 4 | 4 | 5 | 6 | 6 |

## ANTILOGARITHMS.

| Logarithms | 0 | 1 | 2 | 3 | 4 | 5 | 6 | 7 | 8 | 9 | 1 | 2 | 3 | 4 | 5 | 6 | 7 | 8 | 9 |
|---|---|---|---|---|---|---|---|---|---|---|---|---|---|---|---|---|---|---|---|
| | | | | | | | | | | | | | | Proportional Parts. | | | | | |
| ·50 | 3162 | 3170 | 3177 | 3184 | 3192 | 3199 | 3206 | 3214 | 3221 | 3228 | 1 | 1 | 2 | 3 | 4 | 4 | 5 | 6 | 7 |
| ·51 | 3236 | 3243 | 3251 | 3258 | 3266 | 3273 | 3281 | 3289 | 3296 | 3304 | 1 | 2 | 2 | 3 | 4 | 5 | 5 | 6 | 7 |
| ·52 | 3311 | 3319 | 3327 | 3334 | 3342 | 3350 | 3357 | 3365 | 3373 | 3381 | 1 | 2 | 2 | 3 | 4 | 5 | 5 | 6 | 7 |
| ·53 | 3388 | 3396 | 3404 | 3412 | 3420 | 3428 | 3436 | 3443 | 3451 | 3459 | 1 | 2 | 2 | 3 | 4 | 5 | 6 | 6 | 7 |
| ·54 | 3467 | 3475 | 3483 | 3491 | 3499 | 3508 | 3516 | 3524 | 3532 | 3540 | 1 | 2 | 2 | 3 | 4 | 5 | 6 | 6 | 7 |
| ·55 | 3548 | 3556 | 3565 | 3573 | 3581 | 3589 | 3597 | 3606 | 3614 | 3622 | 1 | 2 | 2 | 3 | 4 | 5 | 6 | 7 | 7 |
| ·56 | 3631 | 3639 | 3648 | 3656 | 3664 | 3673 | 3681 | 3690 | 3698 | 3707 | 1 | 2 | 3 | 3 | 4 | 5 | 6 | 7 | 8 |
| ·57 | 3715 | 3724 | 3733 | 3741 | 3750 | 3758 | 3767 | 3776 | 3784 | 3793 | 1 | 2 | 3 | 3 | 4 | 5 | 6 | 7 | 8 |
| ·58 | 3802 | 3811 | 3819 | 3828 | 3837 | 3846 | 3855 | 3864 | 3873 | 3882 | 1 | 2 | 3 | 4 | 4 | 5 | 6 | 7 | 8 |
| ·59 | 3890 | 3899 | 3908 | 3917 | 3926 | 3936 | 3945 | 3954 | 3963 | 3972 | 1 | 2 | 3 | 4 | 5 | 5 | 6 | 7 | 8 |
| ·60 | 3981 | 3990 | 3999 | 4009 | 4018 | 4027 | 4036 | 4046 | 4055 | 4064 | 1 | 2 | 3 | 4 | 5 | 6 | 6 | 7 | 8 |
| ·61 | 4074 | 4083 | 4093 | 4102 | 4111 | 4121 | 4130 | 4140 | 4150 | 4159 | 1 | 2 | 3 | 4 | 5 | 6 | 7 | 8 | 9 |
| ·62 | 4169 | 4178 | 4188 | 4198 | 4207 | 4217 | 4227 | 4236 | 4246 | 4256 | 1 | 2 | 3 | 4 | 5 | 6 | 7 | 8 | 9 |
| ·63 | 4266 | 4276 | 4285 | 4295 | 4305 | 4315 | 4325 | 4335 | 4345 | 4355 | 1 | 2 | 3 | 4 | 5 | 6 | 7 | 8 | 9 |
| ·64 | 4365 | 4375 | 4385 | 4395 | 4406 | 4416 | 4426 | 4436 | 4446 | 4457 | 1 | 2 | 3 | 4 | 5 | 6 | 7 | 8 | 9 |
| ·65 | 4467 | 4477 | 4487 | 4498 | 4508 | 4519 | 4529 | 4539 | 4550 | 4560 | 1 | 2 | 3 | 4 | 5 | 6 | 7 | 8 | 9 |
| ·66 | 4571 | 4581 | 4592 | 4603 | 4613 | 4624 | 4634 | 4645 | 4656 | 4667 | 1 | 2 | 3 | 4 | 5 | 6 | 7 | 9 | 10 |
| ·67 | 4677 | 4688 | 4699 | 4710 | 4721 | 4732 | 4742 | 4753 | 4764 | 4775 | 1 | 2 | 3 | 4 | 5 | 7 | 8 | 9 | 10 |
| ·68 | 4786 | 4797 | 4808 | 4819 | 4831 | 4842 | 4853 | 4864 | 4875 | 4887 | 1 | 2 | 3 | 4 | 6 | 7 | 8 | 9 | 10 |
| ·69 | 4898 | 4909 | 4920 | 4932 | 4943 | 4955 | 4966 | 4977 | 4989 | 5000 | 1 | 2 | 3 | 5 | 6 | 7 | 8 | 9 | 10 |
| ·70 | 5012 | 5023 | 5035 | 5047 | 5058 | 5070 | 5082 | 5093 | 5105 | 5117 | 1 | 2 | 4 | 5 | 6 | 7 | 8 | 9 | 11 |
| ·71 | 5129 | 5140 | 5152 | 5164 | 5176 | 5188 | 5200 | 5212 | 5224 | 5236 | 1 | 2 | 4 | 5 | 6 | 7 | 8 | 10 | 11 |
| ·72 | 5248 | 5260 | 5272 | 5284 | 5297 | 5309 | 5321 | 5333 | 5346 | 5358 | 1 | 2 | 4 | 5 | 6 | 7 | 9 | 10 | 11 |
| ·73 | 5370 | 5383 | 5395 | 5408 | 5420 | 5433 | 5445 | 5458 | 5470 | 5483 | 1 | 3 | 4 | 5 | 6 | 8 | 9 | 10 | 11 |
| ·74 | 5495 | 5508 | 5521 | 5534 | 5546 | 5559 | 5572 | 5585 | 5598 | 5610 | 1 | 3 | 4 | 5 | 6 | 8 | 9 | 10 | 12 |
| ·75 | 5623 | 5636 | 5649 | 5662 | 5675 | 5689 | 5702 | 5715 | 5728 | 5741 | 1 | 3 | 4 | 5 | 7 | 8 | 9 | 10 | 12 |
| ·76 | 5754 | 5768 | 5781 | 5794 | 5808 | 5821 | 5834 | 5848 | 5861 | 5875 | 1 | 3 | 4 | 5 | 7 | 8 | 9 | 11 | 12 |
| ·77 | 5888 | 5902 | 5916 | 5929 | 5943 | 5957 | 5970 | 5984 | 5998 | 6012 | 1 | 3 | 4 | 5 | 7 | 8 | 10 | 11 | 12 |
| ·78 | 6026 | 6039 | 6053 | 6067 | 6081 | 6095 | 6109 | 6124 | 6138 | 6152 | 1 | 3 | 4 | 6 | 7 | 8 | 10 | 11 | 13 |
| ·79 | 6166 | 6180 | 6194 | 6209 | 6223 | 6237 | 6252 | 6266 | 6281 | 6295 | 1 | 3 | 4 | 6 | 7 | 9 | 10 | 11 | 13 |
| ·80 | 6310 | 6324 | 6339 | 6353 | 6368 | 6383 | 6397 | 6412 | 6427 | 6442 | 1 | 3 | 4 | 6 | 7 | 9 | 10 | 12 | 13 |
| ·81 | 6457 | 6471 | 6486 | 6501 | 6516 | 6531 | 6546 | 6561 | 6577 | 6592 | 2 | 3 | 5 | 6 | 8 | 9 | 11 | 12 | 14 |
| ·82 | 6607 | 6622 | 6637 | 6653 | 6668 | 6683 | 6699 | 6714 | 6730 | 6745 | 2 | 3 | 5 | 6 | 8 | 9 | 11 | 12 | 14 |
| ·83 | 6761 | 6776 | 6792 | 6808 | 6823 | 6839 | 6855 | 6871 | 6887 | 6902 | 2 | 3 | 5 | 6 | 8 | 9 | 11 | 13 | 14 |
| ·84 | 6918 | 6934 | 6950 | 6966 | 6982 | 6998 | 7015 | 7031 | 7047 | 7063 | 2 | 3 | 5 | 6 | 8 | 10 | 11 | 13 | 15 |
| ·85 | 7079 | 7096 | 7112 | 7129 | 7145 | 7161 | 7178 | 7194 | 7211 | 7228 | 2 | 3 | 5 | 7 | 8 | 10 | 12 | 13 | 15 |
| ·86 | 7244 | 7261 | 7278 | 7295 | 7311 | 7328 | 7345 | 7362 | 7379 | 7396 | 2 | 3 | 5 | 7 | 8 | 10 | 12 | 13 | 15 |
| ·87 | 7413 | 7430 | 7447 | 7464 | 7482 | 7499 | 7516 | 7534 | 7551 | 7568 | 2 | 3 | 5 | 7 | 9 | 10 | 12 | 14 | 16 |
| ·88 | 7586 | 7603 | 7621 | 7638 | 7656 | 7674 | 7691 | 7709 | 7727 | 7745 | 2 | 4 | 5 | 7 | 9 | 11 | 12 | 14 | 16 |
| ·89 | 7762 | 7780 | 7798 | 7816 | 7834 | 7852 | 7870 | 7889 | 7907 | 7925 | 2 | 4 | 5 | 7 | 9 | 11 | 13 | 14 | 16 |
| ·90 | 7943 | 7962 | 7980 | 7998 | 8017 | 8035 | 8054 | 8072 | 8091 | 8110 | 2 | 4 | 6 | 7 | 9 | 11 | 13 | 15 | 17 |
| ·91 | 8128 | 8147 | 8166 | 8185 | 8204 | 8222 | 8241 | 8260 | 8279 | 8299 | 2 | 4 | 6 | 8 | 9 | 11 | 13 | 15 | 17 |
| ·92 | 8318 | 8337 | 8356 | 8375 | 8395 | 8414 | 8433 | 8453 | 8472 | 8492 | 2 | 4 | 6 | 8 | 10 | 12 | 14 | 15 | 17 |
| ·93 | 8511 | 8531 | 8551 | 8570 | 8590 | 8610 | 8630 | 8650 | 8670 | 8690 | 2 | 4 | 6 | 8 | 10 | 12 | 14 | 16 | 18 |
| ·94 | 8710 | 8730 | 8750 | 8770 | 8790 | 8810 | 8831 | 8851 | 8872 | 8892 | 2 | 4 | 6 | 8 | 10 | 12 | 14 | 16 | 18 |
| ·95 | 8913 | 8933 | 8954 | 8974 | 8995 | 9016 | 9036 | 9057 | 9078 | 9099 | 2 | 4 | 6 | 8 | 10 | 12 | 15 | 17 | 19 |
| ·96 | 9120 | 9141 | 9162 | 9183 | 9204 | 9226 | 9247 | 9268 | 9290 | 9311 | 2 | 4 | 6 | 8 | 11 | 13 | 15 | 17 | 19 |
| ·97 | 9333 | 9354 | 9376 | 9397 | 9419 | 9441 | 9462 | 9484 | 9506 | 9528 | 2 | 4 | 7 | 9 | 11 | 13 | 15 | 17 | 20 |
| ·98 | 9550 | 9572 | 9594 | 9616 | 9638 | 9661 | 9683 | 9705 | 9727 | 9750 | 2 | 4 | 7 | 9 | 11 | 13 | 16 | 18 | 20 |
| ·99 | 9772 | 9795 | 9817 | 9840 | 9863 | 9886 | 9908 | 9931 | 9954 | 9977 | 2 | 5 | 7 | 9 | 11 | 14 | 16 | 18 | 20 |

# INDEX.

*The numbers refer to pages.*

# 218      INDEX.

CAMBRIDGE: PRINTED BY C. J. CLAY, M.A. & SONS, AT THE UNIVERSITY PRESS.

www.ingramcontent.com/pod-product-compliance
Lightning Source LLC
Chambersburg PA
CBHW021657210326
41599CB00013B/1453